DNA Clo

The Practical Approach Series

SERIES EDITORS

D. RICKWOOD
Department of Biology, University of Essex
Wivenhoe Park, Colchester, Essex CO4 3SQ, UK

B. D. HAMES
Department of Biochemistry and Molecular Biology
University of Leeds, Leeds LS2 9JT, UK

★ indicates new and forthcoming titles

Affinity Chromatography
Anaerobic Microbiology
Animal Cell Culture (2nd Edition)
Animal Virus Pathogenesis
Antibodies I and II
★ Basic Cell Culture
Behavioural Neuroscience
Biochemical Toxicology
★ Bioenergetics
Biological Data Analysis
Biological Membranes
Biomechanics – Materials
Biomechanics – Structures and Systems
Biosensors
★ Carbohydrate Analysis (2nd Edition)
Cell–Cell Interactions
★ The Cell Cycle
★ Cell Growth and Apoptosis
Cellular Calcium
Cellular Interactions in Development
Cellular Neurobiology
Clinical Immunology
Crystallization of Nucleic Acids and Proteins
★ Cytokines (2nd Edition)
The Cytoskeleton
Diagnostic Molecular Pathology I and II
Directed Mutagenesis
★ DNA Cloning 1: Core Techniques (2nd Edition)
★ DNA Cloning 2: Expression Systems (2nd Edition)
★ DNA Cloning 3: Complex Genomes (2nd Edition)
★ DNA Cloning 4: Mammalian Systems (2nd Edition)
Electron Microscopy in Biology
Electron Microscopy in Molecular Biology
Electrophysiology

Enzyme Assays
Essential Developmental Biology
Essential Molecular Biology I and II
Experimental Neuroanatomy
★ Extracellular Matrix
★ Flow Cytometry (2nd Edition)
Gas Chromatography
Gel Electrophoresis of Nucleic Acids (2nd Edition)
Gel Electrophoresis of Proteins (2nd Edition)
★ Gene Probes 1 and 2
Gene Targeting
Gene Transcription
Glycobiology
Growth Factors
Haemopoiesis
Histocompatibility Testing
★ HIV Volumes 1 and 2
Human Cytogenetics I and II (2nd Edition)
Human Genetic Disease Analysis
Immunocytochemistry
In Situ Hybridization
Iodinated Density Gradient Media
★ Ion Channels
Lipid Analysis
Lipid Modification of Proteins
Lipoprotein Analysis
Liposomes
Mammalian Cell Biotechnology
Medical Bacteriology
Medical Mycology
★ Medical Parasitology
★ Medical Virology
Microcomputers in Biology
Molecular Genetic Analysis of Populations
★ Molecular Genetics of Yeast
Molecular Imaging in Neuroscience
Molecular Neurobiology
Molecular Plant Pathology I and II
Molecular Virology
Monitoring Neuronal Activity
Mutagenicity Testing
Neural Transplantation
Neurochemistry
Neuronal Cell Lines
NMR of Biological Macromolecules
★ Non-isotopic Methods in Molecular Biology
Nucleic Acid Hybridization
Nucleic Acid and Protein Sequence Analysis
Oligonucleotides and Analogues
Oligonucleotide Synthesis
PCR 1
★ PCR 2
★ Peptide Antigens
Photosynthesis: Energy Transduction
★ Plant Cell Biology
★ Plant Cell Culture (2nd Edition)
Plant Molecular Biology
★ Plasmids (2nd Edition)

Pollination Ecology
Postimplantation Mammalian Embryos
Preparative Centrifugation
Prostaglandins and Related Substances
★ Protein Blotting
Protein Engineering
Protein Function
Protein Phosphorylation
Protein Purification Applications
Protein Purification Methods
Protein Sequencing
Protein Structure
Protein Targeting
Proteolytic Enzymes
★ Pulsed Field Gel Electrophoresis
Radioisotopes in Biology
Receptor Biochemistry
Receptor–Effector Coupling
Receptor–Ligand Interactions
RNA Processing I and II
Signal Transduction
Solid Phase Peptide Synthesis
Transcription Factors
Transcription and Translation
Tumour Immunobiology
Virology
Yeast

DNA Cloning 3
A Practical Approach

Edited by

D. M. GLOVER

CRC Laboratories
Department of Anatomy and Physiology
University of Dundee

and

B. D. HAMES

Department of Biochemistry and Molecular Biology,
University of Leeds

Walton Street, Oxford OX2 6DP
d New York
nd Bangkok Bombay
vn Dar es Salaam Delhi
Kong Istanbul Karachi
dras Madrid Melbourne
Mexico City Nairobi Paris Singapore
Taipei Tokyo Toronto
and associated companies in
Berlin Ibadan

Oxford is a trade mark of Oxford University Press

Published in the United States
by Oxford University Press Inc., New York

A catalogue record for this book is available from the British Library

Library of Congress Cataloging-in-Publication Data
DNA cloning 3 : a practical approach / edited by D.M. Glover and B.D. Hames.
(A Practical approach series; no. 163)
Includes bibliographical references and index.
1. Molecular cloning—Laboratory manuals. 2. Recombinant DNA—Laboratory manuals. 3. Molecular cloning—Laboratory manuals. 4. Recombinant DNA—Laboratory manuals. I. Glover, David M. II. Hames, B. D. III. Series.
QH442.2.D63 1995 574.87′3282—dc20 95-14776

ISBN 0 19 963482 3 (Pbk)
ISBN 0 19 963483 1 (Hbk)

Typeset by Footnote Graphics, Warminster, Wilts
Printed in Great Britain by Information Press Ltd, Eynsham, Oxon.

Preface

It is a decade since the first editions of *DNA Cloning* were being prepared for the Practical Approach series. It is illuminating to look back at those volumes and reflect how the field has evolved over that period. We have tried to distil out of those earlier volumes such techniques that have withstood the test of time, and have asked many of the former contributors to update their chapters. We have also invited several new authors to write chapters in areas that have come to the forefront of this invaluable technology over recent years. The field is, however, far too large to cover comprehensively, and so we have had to be selective in the areas we have chosen. This has also led to each of the books being focused onto particular topics. This having been said, Volume 1 covers core techniques that are central to the cloning and analysis of DNA in most laboratories. Volume 2 on the other hand turns to the systems used for expressing cloned genes. Inevitably the descriptions of these techniques can be supplemented by reference to other cloning manuals such as *Molecular cloning* by Sambrook, Fritsch, and Maniatis (Cold Spring Harbor Laboratory, New York, 1989) as well as other books in the Practical Approach series. Volume 3 examines the analysis of complex genomes, an area in which there have been many important developments in recent years, both in the description of new vectors and in the strategic approaches to genome mapping. Again, companion volumes such as *Genome analysis* can be found in the Practical Approach series. Finally, in Volume 4 we look at DNA cloning and expression in mammalian systems; from cultured cells to the whole animal.

Volume 3 of the first edition of *DNA Cloning* was published in 1987, and of that very little remains in this edition. Its contents have been divided into the companion books on 'Expression Systems', 'Complex Genomes', and 'Mammalian Systems', and each has been expanded in its scope. Chapters 1 and 2 of this volume had their counterparts in the earlier edition. These chapters have retained their senior authors and give an up-to-date description of how best to carry out genome analysis using cosmids almost a decade later. In the intervening time, other vector systems for cloning large segments of DNA have come into their own. Of these, perhaps the most successful have been P1 phage covered by Nat Sternberg in Chapter 3, and yeast artificial chromosomes (YACs) described by Rakesh Anand in Chapter 4. The development of PCR in this period of time has resulted in the procedure for obtaining chromosomal material by microdissection becoming more readily available to a wider group of researchers. Protocols for carrying out this approach either with Dipteran polytene chromosomes, or with mitotic chromosomes are described by Robert Saunders in Chapter 5. Genome analysis now requires an interplay of molecular biology and computational approaches. Although this is an area covered more extensively in other books within this series, we felt it useful to include some central techniques in this volume. This has led to a discussion of how to access and use DNA sequence databases by Rainer Fuchs and Graham Cameron in Chapter 6. Finally, Wendy Bickmore describes techniques for long-range restriction mapping in Chapter 7, and Isam Naom, Christopher Matthew, and Margaret-Mary Town cover the powerful techniques of constructing genetic maps using microsatellite polymorphisms in Chapter 8.

As with its sister volumes, we hope that this book will find its way onto bookshelves in laboratories, and in so doing will become messy and 'dog-eared'. It has been gratifying to see how widespread the use of the first edition has become, and a similar success would be rewarding to both the editors and the authors. Finally and most importantly we would like to thank all of the authors for their contributions and for their patience in bringing this project to fruition.

Dundee and Leeds David M. Glover
May 1995 B. David Hames

Contents

Contributors

RAKESH ANAND
Genome Group, Zeneca Pharmaceuticals, Alderley Park, Macclesfield, Cheshire SK10 4TG, UK.

WENDY BICKMORE
MRC Human Genetics Unit, Western General Hospital, Crewe Road, Edinburgh EH4 2XU, UK.

GRAHAM N. CAMERON
EMBL Data Library, European Molecular Biology Laboratory, Postfach 10.2209, 69012 Heidelberg, Germany.

RAINER FUCHS
Bioinformatics Group, Glaxo Research Institute, 5 Moore Drive, Research Triangle Park, NC 27709, USA.

ALASDAIR C. IVENS
Department of Biochemistry, Imperial College of Science, Technology and Medicine, Exhibition Road, London SW7 2AZ, UK.

HANS LEHRACH
Genome Analysis Department, Imperial Cancer Research Fund, 44 Lincoln's Inn Fields, London WC2A 3PX, UK.

PETER F. R. LITTLE
Department of Biochemistry, Imperial College of Science, Technology and Medicine, Exhibition Road, London SW7 2AZ, UK.

CHRISTOPHER G. MATHEW
Division of Medical and Molecular Genetics, United Medical and Dental Schools of Guy's and St. Thomas's Hospitals, Guy's Campus, London SE1 9RT, UK.

ISAM S. NAOM
Division of Medical and Molecular Genetics, United Medical and Dental Schools of Guy's and St. Thomas's Hospitals, Guy's Campus, London SE1 9RT, UK.

DEAN NIŽETIĆ
Centre for Applied Molecular Biology, University of London, School of Pharmacy, 29/39 Brunswick Square, London WC1N 1AX, UK.

ROBERT D. C. SAUNDERS
Department of Anatomy and Physiology, University of Dundee, Dundee DD1 4HN, UK.

NAT STERNBERG
Dupont-Merck Pharmaceutical Co., Glenolden Laboratories, 500 S. Ridgeway Avenue, Glenolden, PA 19036, USA.

MARGARET-MARY TOWN
Division of Medical and Molecular Genetics, United Medical and Dental Schools of Guy's and St. Thomas's Hospitals, Guy's Campus, London SE1 9RT, UK.

Abbreviations

BAC	bacterial artificial chromosome
bp	base pairs
BSA	bovine serum albumin
c.f.u.	colony-forming units
CHEF	contour-clamped homogeneous field electrophoresis
CIP	calf intestinal phosphatase
cM	centimorgan
CMGT	cell-mediated gene transfer
DAB	diaminobenzidine
DTT	dithiothreitol
EDTA	ethylenediaminetetraacetic acid
FIGE	field inversion gel electrophoresis
FISH	fluorescence *in situ* hybridization
FTP	file transfer protocol
GDP	genome database
HRP	horse-radish peroxidase
kb	kilobase
MSP	microsatellite polymorphisms
NTA	nitrilo-triacetic acid
OPC	oligonucleotide purification column
PCR	polymerase chain reaction
PFGE	pulsed field gel electrophoresis
PMSF	phenylmethylsulfonyl fluoride
RFLP	restriction fragment length polymorphism
SDS	sodium dodecyl sulfate
SRS	sequence retrieval system
STS	sequence tag site
TBE	Tris, borate, EDTA
TE	Tris, EDTA
TEAA	triethylammonium acetate
TEMED	*N,N,N',N'*-tetramethylethylenediamine
VRC	vanadyl ribonucleoside complexes
WAIS	wide area information system
WWW	world-wide web
YAC	yeast artificial chromosome

1

Cosmid clones and their application to genome studies

ALASDAIR C. IVENS and PETER F. R. LITTLE

1. Introduction: scope of the chapter

A 1000-fold range of DNA sizes may be cloned in the current range of cloning vectors: the ideal vector for genome mapping studies would be one that is easy to use while containing as much DNA as possible. Cosmids now occupy the middle ground: they have a significant capacity, thus reducing the number of steps required to clone an entire gene or region, combined with very simple methods for isolating inserted DNA in pure form. As a consequence, positional cloning strategies frequently involve the use of cosmids as a final cloning vector for reducing a yeast artificial chromosome (YAC) clone to manageably sized DNA fragments (1). Cosmids are units that are very likely to contain an entire gene, are easily mapped with respect to restriction sites, and are amenable to the application of a number of other functional assays, e.g. exon trapping (2, 3), genomic sequencing (4–6), and fingerprinting to generate contig maps (7, 8). As a result, the detailed information that can be obtained from a cosmid clone makes it the ideal medium for genome analysis. Indeed, several genome mapping studies (e.g. *Caenorhabditis elegans* (9), *Escherichia coli* (10)), have relied on physical DNA maps built around cosmid clones.

Cosmids are likely to maintain this middle ground for the foreseeable future. P1 (11) and bacterial artificial chromosome (BAC) vectors (12) are the most likely competitors, but paradoxically, their inserts may be too large for some sorts of analyses. Restriction enzyme digestion of the 100–250 kilobase (kb) DNAs carried by these vectors commonly generates a slew of fragments that precludes simple gel isolation. The smaller inserts in cosmid vectors, combined with the relative ease of preparation compared to P1s or BACs, will remain a significant advantage.

Genome mapping projects, until relatively recently, have focused almost exclusively on specific genetic and regional targets. Whatever the method used to isolate the region of interest (e.g. cell-mediated gene transfer (CMGT), radiation hybrids, YAC cloning, or pulsed field gel electrophoresis (PFGE)), cosmid cloning has frequently been used for further analysis. A characteristic of some of these techniques is that they generate fragments of DNA from one species (transgenomes) on the background of the complete genome of another species. The construction of cosmid libraries, and the subsequent generation of a linear map of the region by fingerprinting and contig analysis, is an important prelude to the detailed dissection of these transgenomes by other methods.

The purpose of this chapter is first to record a series of protocols for constructing cosmid libraries that have worked in our laboratory. Secondly, this chapter deals with some of the subsequent analyses available for cosmids derived from a region of interest: specifically, we describe clone fingerprinting. The protocols detailed below should be treated as guide-lines only. We would, however, advise that inexperienced users do not deviate from them until enough experience has been attained.

2. Construction of cosmid libraries

2.1 Cosmid vectors: a historical perspective

Cosmid vectors were first developed by Collins and Hohn (13) to overcome the technical problems of introducing large pieces of DNA into *E. coli*. Most protocols for the preparation of competent *E. coli* allow efficient transfer of DNA molecules up to approximately 10–15 kb and attempts to use DNA larger than this result in a dramatic reduction of transformation efficiencies. This remains true today, despite the use of electroporation techniques. The major barrier to successful incorporation of large plasmids into bacterial cells appears to be in the transfer of DNA across the cell membrane, or perhaps in the events immediately following entry of the DNA into the periphery of the cytoplasm.

To overcome these problems, Hohn and Collins (14) introduced the bacteriophage λ *cos* sequence into a conventional plasmid cloning vector. This allowed the use of λ phage *in vitro* packaging protocols (15, 16) which completely overcomes the size effects seen in transfection. *Cos*, and approximately 200 bp of DNA either side, is a DNA sequence that is required for packaging DNA into preformed λ phage particles. No other specific λ DNA segments are necessary. For successful packaging of DNA, there must be two *cos* sequences contained on the same DNA molecule, separated by 38–52 kb, which are the packaging limits of lambda. Generation of recombinant cosmid molecules is made possible by ligating 30–45 kb of, say, human DNA to suitably prepared cosmid vector DNA in ligation reactions that favour the formation of long concatemers of cosmid/human/cosmid DNA. This creates the necessary distribution of *cos* sites that can be used in *in vitro* packaging reactions, with biological properties almost identical to the *in vivo* process. *In vitro* packaging can be remarkably efficient with up to 10^9 plaques formed per microgram of λ DNA. However, this corresponds to only a 5% efficiency with respect to the number of DNA molecules, and the efficiency of cosmid transduction is much lower: rarely greater than 10^6 colony-forming units (c.f.u.) per microgram, and generally considerably less.

2.2 Choice of vector

The first criterion for choosing a cosmid vector must be to establish whether a cosmid-size clone is actually required. If this is not the case, then construction of phage libraries is often easier and quicker. There are many different cosmid vectors available (see refs 17 and 18 for listings). Important factors that need to be considered are drug resistance, the replicon, and the positioning of convenient restriction sites that will facilitate subsequent analysis. These are discussed below. The ultimate decision about which vector to use will probably be a compromise.

Most cosmids that are suitable for use in *E. coli* are derived from pBR322 and based upon the pMBI replicon. The control of this replicon is not ideal for use in a cosmid vector (discussed in ref. 19) and can lead to overgrowth of libraries by small cosmids, derived by *recA*-independent deletion events. There may also be problems with interference of the replicon by adventitious promoters in the insert DNA (see ref. 20). The use of the phage λ origin of replication has overcome some, but not all, of these problems (21, 22).

Cosmid-containing cells often grow slowly for reasons that are complex and not well understood. Slow growth is often associated with partial resistance to commonly used antibiotics (e.g. ampicillin) which may reduce selection pressure on the cosmid. The 'satellite' colonies that result may also be a problem in the initial slow phase of cosmid growth. While many cosmid vectors contain this drug resistance marker, it is not the optimal drug resistance determinant. Tetracycline resistance does not suffer from 'satellite' formation, but the drug is significantly unstable, being light-sensitive. Kanamycin (or neomycin) resistance, which can be encoded for by two different genes, is used in several vectors and is a stable drug. Its major disadvantage is that long (up to one hour) pre-expression times are required to allow full expression, though this is a small price to pay in light of the other advantages.

When choosing a cosmid vector, a major consideration should naturally be the presence of the appropriate cloning site. Additional points worth considering are the presence of sites flanking the cloning site(s) that enable cleavage of the insert DNA away from the vector and the presence of sites between pairs of *cos* sequences that allows even more simple generation of cosmid arms (23). Generally, cloning strategies follow the conventional strategy of *Sau*3AI partial digests inserted into a *Bam*HI site as shown in *Figure 1*.

Many *Bam*HI vectors are available, e.g. LoristB (20). However, the analysis of such clones for identification of cosmid overlaps, for example, is difficult since the ends of the insert DNA are lost to the analysis—the fusion of a *Sau*3AI and *Bam*HI site destroys the *Bam*HI site 75% of the time. Partial

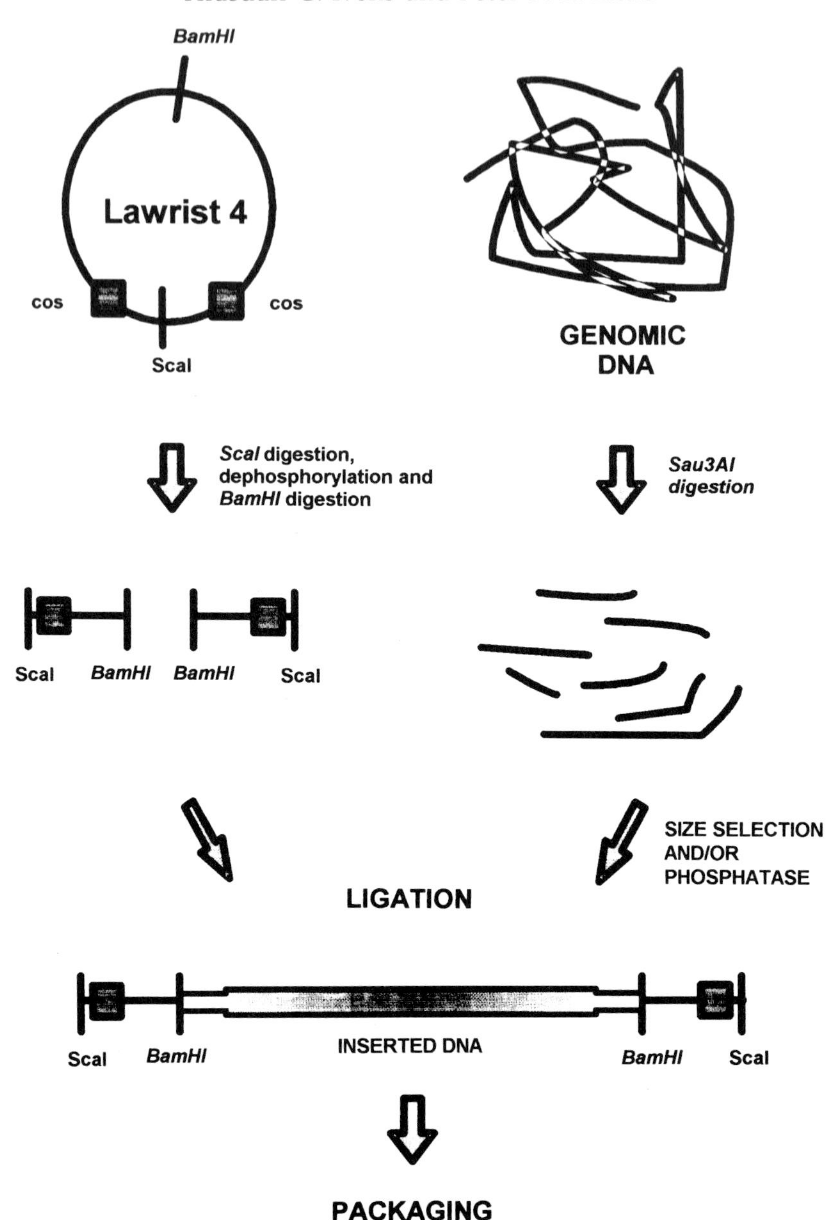

Figure 1. Cosmid cloning procedure. The two parallel procedures are shown schematically. (**Note**: not to scale.) On the left-hand side of the diagram, Lawrist4 vector is cleaved at the unique *Sca*I site, dephosphorylated, and subsequently digested with *Bam*HI to give two fragments or 'arms', each of which contains a *cos* site. On the right-hand side of the diagram, high molecular weight genomic DNA is partially digested with *Sau*3AI, and size selected to give fragments of approximately 40–50 kb in length. The two components are ligated together, and packaged into phage particles suitable for transfection into *E. coli*.

digests generated by restriction enzymes with a six base recognition sequence overcomes this problem. The representation of libraries made in this fashion, however, is theoretically not very good (24).

We have made extensive use of lambda origin cosmid vectors (20–22), in particular the double *cos* Lawrist series of vectors, a derivative Lorist engineered by Pieter de Jong (personal communication). These vectors, which increase the overall efficiency of cloning by both facilitating vector preparation and increasing packaging efficiency, have a variety of cloning and manipulation sites and can be used in the Bates and Swift (23) protocols. The most widely used vector is Lawrist4 which, upon packaging, yields the LoristX vector (25) which is a minor modification of LoristB (20).

The Lawrist vectors were made by recombining a *cos*-containing pUC derivative into LoristX, yielding a double *cos* vector that replicates with the pUC copy number. Recombination is reversible, however; non-recombinant Lawrist vectors will always have a small proportion (visible on gel electrophoresis) of the pUC 'cosmid', though this can be ignored as it has no cleavage sites for the restriction enzymes used in the cloning procedure. Once packaged, the vector is LoristX and completely stable. Yields of the Lawrist series of vectors are high and simple to manipulate. Cosmid clones generated using this series of vectors have been used in a wide range of large scale genome mapping projects (9, 26).

Another vector that might warrant consideration is pWE15 (27). This dual-*cos* cosmid vector incorporates restriction enzyme 'cassettes' around the cloning site that enable both simple removal of the inserted DNA and mapping/sequencing of the insert DNA. On the 'negative' side, this vector confers ampicillin resistance to the host cell, and is a pMB1 replicon.

2.3 Preparation of genomic DNA for cloning

The protocols described throughout the chapter utilize a common set of reagents that are given in *Table 1*. In this section, we describe several approaches for preparing high molecular weight DNA for cloning (*Protocols 1, 2*, and *3*), and its partial digestion (*Protocols 4* and *5*). There are two basic

Table 1. General reagents used in many protocols

99% and 70% ethanol
Propan-2-ol (isopropanol)
TE: 10 mM Tris–HCl, 1 mM EDTA, pH 8.0
10 × TBE: 106 g Tris base, 55 g boric acid, 9.5 g disodium EDTA per litre, pH 9.3
20 × SSC: 3 M NaCl, 0.3 M Na citrate pH 7.0
Ethidium bromide (10 mg/ml)
High gelling temperature agarose (IBI, IB70042)

protocols available for preparing eukaryotic DNA of a suitable size range for cloning:

- physical separation of partially digested fragments of the correct size (*Protocol 6*)
- dephosphorylation of the partial digest, relying on the natural size selectivity of *in vitro* packaging to isolate 35–45 kb fragments (*Protocol 7*)

The former is expensive on starting material: 300 μg of DNA will be required to yield sufficient DNA to construct a library with high efficiency (approximately 10^6 colonies per microgram of eukaryotic DNA). The dephosphorylation method is more economic in its use of starting DNA but can be less efficient.

The quality of the DNA used to make a cosmid library is perhaps the most critical component of the entire library construction procedure. We use the method detailed in *Protocol 1* to make high molecular weight DNA from tissues or cell lines. The starting DNA must be of very high quality. DNA, analysed by agarose gel electrophoresis, should ideally be greater than 200 kb in size, though greater than 150 kb is usually acceptable. We have never obtained a representative library with DNA smaller than this: only approximately one-third of partially digested DNA molecules of size range 40–50 kb derived from molecules of starting size 100 kb have restriction sites at both ends. The broken molecules compete in the ligation reaction and make a successful library virtually impossible to construct. The most important advice we can offer is that the DNA should be checked at every stage of the isolation (partials, gradients, etc.) by electrophoresis through either 0.2% agarose gels, or by field inversion gel electrophoresis (FIGE). Instructions for these are included in *Protocols 17* and *18*. RNase digestion and ethanol precipitation/resuspension also reduce the molecular weight of the DNA, presumably by shearing, and are to be avoided if possible.

Protocol 1. Preparation of eukaryotic DNA

Reagents

- Proteinase K (Boehringer Mannheim, 1092766)
- TNE: 10 mM Tris–HCl pH 8.0, 100 mM NaCl, 1 mM EDTA
- TNES: 100 mM Tris–HCl pH 8.0, 100 mM NaCl, 10 mM EDTA, 1% Sarkosyl pH 7.5

Method

1. Resuspend or homogenize approximately 10^8–10^9 cells (or 0.5–1 g tissue) in 5 ml of 1 × SSC, making sure resuspension is complete.
2. Add 5 ml of TNES. Mix carefully for approximately 2–3 min to lyse the cells. Add proteinase K to 100 μg/ml. Incubate at 55°C for 2 h.
3. Extract once with phenol, by adding an equal volume of phenol saturated

with TE. Mix by gently inverting the covered tube and separate the phases by centrifugation. Remove the aqueous phase into a clean tube and carry out a further extraction with an equal volume of phenol/chloroform (1:1). Finally extract once with an equal volume of chloroform alone. If too many cells have been used, the first phenol extraction will be virtually solid. If this has happened dilute the DNA with TNES. The first phenol extraction is often messy but the interface should be as far as possible left with the phenol phase. The subsequent phenol/chloroform and chloroform extractions will be much cleaner.

4. Dialyse the final aqueous phase against 4 litres of TNE at 4°C for 17 h overnight or longer.
5. Repeat the dialysis for 24 h against 4 litres of TE at 4°C.
6. The DNA is now ready for use. The concentration must be estimated by agarose gel electrophoresis against known concentrations of λ DNA. (As the preparation will still contain RNA, a reading of UV absorption at 260 nm will give an inaccurate measure of DNA concentration.) Typically, the DNA will be at a concentration of 100–300 μg/ml.

2.4 Recombinant YAC clones as a starting material for cosmid cloning

Often the DNA region of interest is contained within one or more YAC clones (see Chapter 4). While fingerprinting techniques that take advantage of the polymerase chain reaction (PCR) can be applied to individual yeast chromosomes (28), further analysis of the region can also be achieved by the construction of a cosmid contig map of the region. This necessitates the construction of cosmid 'mini-libraries'. It is possible to construct cosmid libraries from individual YAC chromosomes isolated from PFGE gels, though technically, this is more difficult as the amounts of DNA involved are often small. It is considerably easier to follow progress of the cloning procedure if the DNA being manipulated is actually visible on a gel. The disadvantage of this approach however is that transgenome-derived cosmid clones will have to be identified by hybridization.

Described below are protocols that yield high quality DNA. Methods for the isolation of DNA from YAC clones are also included (*Protocols 2* and *3*). *Protocol 2* results in DNA in solution, while *Protocol 3* provides a method for the preparation of yeast DNA in agarose blocks that is suitable for both pulsed field gel electrophoresis (PFGE) and cosmid libraries. The latter protocol should be scaled up as deemed necessary. Similar protocols are given in Chapter 4 (e.g. *Protocol 8*). The variations between these protocols reflect differences in practise between laboratories. The reader is encouraged to experiment with these variations as skill is acquired.

Protocol 2. Preparation of yeast DNA in solution

Reagents

- 1 × amino acid supplement: 20 μg/ml adenine, 20 μg/ml arginine, 20 μg/ml isoleucine, 20 μg/ml histidine, 60 μg/ml leucine, 20 μg/ml lysine, 20 μg/ml methionine, 50 μg/ml phenylalanine, 0.15 mg/ml valine, 30 μg/ml tyrosine (Sigma)
- SCE: 1 M sorbitol, 0.1 M sodium citrate, 10 mM EDTA, 14 mM 2-mercaptoethanol, pH 5.8
- Yeast medium: 1.3% yeast nitrogen base, 4% glucose, 1 × amino acid supplement
- Zymolase: 10 mg/ml in 10 mM Tris–HCl pH 7.5 (ICN Biomedicals, 320931)
- TNE: 10 mM Tris–HCl pH 8.0, 100 mM NaCl, 1 mM EDTA
- Proteinase K (Boehringer Mannheim, 1092766)

Method

1. Culture the YAC-containing yeast cells at 30°C with vigorous shaking for 48–72 h in 500 ml yeast medium.
2. Pellet the cells by centrifugation (5000 *g* for 10 min), and resuspend in 30 ml SCE.
3. Add 200 μl 10 mg/ml zymolase and incubate with shaking at 37°C for 1 h.
4. Pellet the resulting spheroblasts by centrifugation at 1000 *g* for 10 min. Resuspend the spheroblasts gently in 25 ml TNE. Add proteinase K to a final concentration of 250 μg/ml and incubate at 55°C for 2 h.
5. Purify the DNA using two phenol/chloroform extractions, followed by a single chloroform extraction. Ensure that all mixing is as gentle as possible to prevent shearing of the DNA.
6. Although it is far preferable to dialyse the extracted DNA against TE pH 8.0 for 24–48 h at 4°C, it is possible to precipitate DNA by the addition of 0.1 vol. 3 M sodium acetate pH 5.5 and 2 vol. of ethanol. Recover the DNA by gentle spooling. Wash the pellet in 70% ethanol, air dry, and resuspend overnight in 500 μl TE.

Protocol 3. Preparation of yeast DNA in agarose blocks [a]

Reagents

- Yeast medium: 1.3% yeast nitrogen base, 4% glucose, 1 × amino acid supplement
- 1 × amino acid supplement: 20 μg/ml adenine, 20 μg/ml arginine, 20 μg/ml isoleucine, 20 μg/ml histidine, 60 μg/ml leucine, 20 μg/ml lysine, 20 μg/ml methionine, 50 μg/ml phenylalanine, 0.15 mg/ml valine, 30 μg/ml tyrosine (Sigma)
- SCE: 1 M sorbitol, 0.1 M sodium citrate, 10 mM EDTA, 14 mM 2-mercaptoethanol, pH 5.8
- Zymolase: 10 mg/ml in 10 mM Tris–HCl pH 7.5 (ICN Biomedicals, 320931)
- Low gelling temperature agarose (FMC Bioproducts, 50102)
- LIDS mix: 1% lithium dodecyl sulfate, 100 mM EDTA, 10 mM Tris–HCl pH 8.0
- TNE: 10 mM Tris–HCl pH 8.0, 100 mM NaCl, 1 mM EDTA
- 1 M sorbitol, 20 mM EDTA pH 8.0

Method

1. Culture the YAC-containing yeast cells in 20 ml yeast medium at 30°C with vigorous shaking for 48–72 h.
2. Dissolve low gelling agarose (2%) in 1 M sorbitol, 20 mM EDTA by boiling. Cool to 50°C and add β-mercaptoethanol to 14 mM. Leave the agarose in a water-bath at 50°C until it is required.
3. Pellet the cells by centrifugation (5000 *g* for 10 min). Wash the cells once in 50 mM EDTA. Re-centrifuge the cells, discard supernatant, and resuspend the pellet in 400 μl SCE.
4. Prepare PFGE block-formers (e.g. Pharmacia, 100 μl per block) on ice, following the manufacturer's instructions.
5. Add sufficient zymolase to the yeast cells to give a final concentration of 100 μg/ml.
6. Add 500 μl of the molten agarose to the yeast cells that have been pre-warmed to 50°C for 1 min, mix quickly, and aliquot into the block formers. Allow to set on ice for 20 min.
7. Carefully transfer the agarose blocks to a 50 ml polypropylene tube and cover the blocks with 5 ml of SCE plus 100 μg/ml zymolase. Incubate at 37°C for 2 h with occasional shaking.
8. Decant the supernatant, and replace with 5 ml filter-sterilized LIDS mix. Incubate at 37°C for 1 h.
9. Repeat the previous step, but leave to incubate at 37°C overnight. (The blocks can be left at this point indefinitely.)
10. Pour off the LIDS mix, and rinse the blocks extensively with TE plus 10 mM EDTA at 50°C. Store at 4°C. Proceed to *Protocol 5* for partial digestions.

[a] See also Chapter 4, *Protocol 2*; Chapter 7, *Protocol 1*.

2.5 Insert DNA preparation

We perform partial digestions with an enzyme dilution series, detailed in *Protocol 4*, adopting the method of Seed *et al.* (24) to identify the correct digestion conditions for optimal library representation. It is also possible to vary the time of digestion to achieve the same end. In this case, aim to add sufficient enzyme such that the digestion should go to completion in 2 h (i.e. 0.5 U enzyme/μg DNA). Take samples at 5 min intervals, and stop the reaction by adding EDTA to a final concentration of 20 mM. Analyse the digestion products by electrophoresis (detailed in *Protocols 17* and *18*). An alternative approach to achieving partial digestion of DNA protected from the restriction endonuclease by methylation is given in Chapter 2, section 2.3.

A critical factor in these procedures is to ensure thorough mixing of the reaction prior to the onset of digestion. This can be difficult if the DNA is of high quality, as it will be extremely viscous. As it is still not clear that all *Sau*3AI restriction sites exhibit the same cleavage kinetics, three different partial digestion conditions should be identified prior to scale up. The method described here is very similar to that described in ref. 17.

Protocol 4. Partial digestion of DNA in solution[a]

Reagents

- *Sau*3AI (Boehringer Mannheim, 709751)

Method

1. Use the DNA, purified as described in *Protocol 1* or *Protocol 2*. Take 180 μl DNA and add 20 μl of the appropriate 10× concentrated restriction endonuclease digestion buffer and mix carefully.
2. Remove a 40 μl aliquot into a tube and pipette 20 μl aliquots of the rest of the sample into eight further tubes. Label these tubes 1–9.
3. Add restriction enzyme to tube 1 containing the 40 μl aliquot. We generally use 2 μl of restriction endonuclease at 10–20 U/μl (where 1 U is that amount of enzyme that will cleave 1 μg λ DNA to completion in 1 h). Mix very carefully and *with a fresh tip* take out 20 μl and add to tube 2. Mix and repeat. Do not add anything to tube 9.
4. Incubate at 37°C for 30 min, then place the tubes on ice.
5. Check the molecular length of the DNA by subjecting 5 μl to electrophoresis on a 1% agarose gel. You should see complete or virtually complete digestion in tube 1 and increasingly partial digestion towards tube 8. If this is observed, analyse 10 μl of the digest by electrophoresis on a 0.2% agarose gel or by FIGE (*Protocols 17* and *18*). Select the correct partial conditions by masking off the DNA in a photograph of the agarose gel that is greater than 50 kb and less than 40 kb. The correct partial is *not* the one with most DNA in this range; it is the next less digested. (The rationale for this is contained in ref. 24.) Use the correct partial conditions and the conditions of the adjacent digests and scale up for the preparative digests.
6. It is important to scale up the reaction to reproduce the above conditions *exactly*. It is necessary to pre-warm the DNA solution since it is a larger volume. Calculate the volume of enzyme required for the appropriate partial digestion conditions. Do not attempt to change buffer, enzyme, or DNA concentration. Aim to partially digest as much DNA as possible. If *Protocol 6* is used for size fractionation, 200–300 μg of DNA will be required.
7. Check the integrity and size of the scaled up digestion products by

electrophoresis on a 0.2% agarose gel (*Protocol 17*) or by FIGE (*Protocol 18*) using suitable size markers.

8. Pool the partial digestion products and extract once with phenol, once with chloroform, and ethanol precipitate the DNA. Resuspend the DNA precipitate in TE to give a concentration of 300 μg/ml.

[a] See also Chapter 2, *Protocol 4*; Chapter 3, *Protocol 2*; Chapter 4, *Protocol 3*.

A method for the partial digestion of DNA within agarose blocks is given in *Protocol 5*.

Protocol 5. Partial digestion of yeast DNA in agarose blocks[a]

Reagents

- *Sau*3AI (Boehringer Mannheim, 709751)
- Calf intestinal phosphatase (Boehringer Mannheim, 1097075)
- Bovine serum albumin (Sigma, B2518)
- Nitrilo-triacetic acid (NTA) (Sigma, N9877)

Method

1. Dialyse the agarose blocks against 50 ml TE for 30 min. Repeat.
2. Set-up a time course to obtain optimal partial digestion conditions, using one 100 μl block per time point. Add 60 μl distilled water, 20 μl 10× *Sau*3AI restriction enzyme buffer, and 10 μl BSA (2 mg/ml) to give a final volume of 190 μl, leaving 10 μl free for the enzyme. Convenient time points are 0, 1, 2.5, 5, 10, 20, and 30 min.
3. Place the tubes at 70°C for 20 min, during which time the agarose will melt.
4. Dilute the restriction enzyme *Sau*3AI in 1× enzyme buffer to give an enzyme concentration of 0.5 U/10 μl.
5. Cool the tubes by placing them in a 37°C water-bath for 5 min. Add 10 μl diluted enzyme to the side of each tube apart from the zero time point, and invert a couple of times to start the reaction. Incubate at 37°C.
6. At the requisite time points, remove the tube to a 70°C water-bath. After the final time point has been reached, also place the zero time point at 70°C. After 20 min, place the tubes at 37°C to cool. Remove 20 μl from each time point for analysis by gel electrophoresis (see *Protocols 17* and *18*). (It is possible to skip the next two steps, proceeding directly to step 9. In this manner, the integrity of the DNA can be tested by self-ligation tests. Once the DNA has been dephosphorylated, this is no longer possible. Assuming self-ligation is observed, the samples **must** subsequently be dephosphorylated as described in *Protocol 7*.)

Protocol 5. *Continued*

7. Add 5 U of calf intestinal phosphatase to each time point, mix carefully, and incubate at 37°C for 20 min.
8. Add 0.1 vol. 0.1 M NTA to the tubes corresponding to each time point and incubate for 20 min at 70°C. Cool to 37°C.
9. Extract the DNA twice with phenol, once with phenol/chloroform, and once with chloroform prior to precipitation with ethanol overnight at −20°C.
10. Recover the DNA by centrifugation, wash the pellet carefully with 70% ethanol, and resuspend in 20 μl TE. Recovery is typically 50–70%. Analyse an aliquot (2 μl) of each time point by electrophoresis, and compare with the result obtained previously (step 6). Choose the suitable time points as described in *Protocol 4*, and proceed to ligation to prepared vector (*Protocol 9*).

[a] See also Chapter 7, *Protocol 5*.

DNA of the appropriate size range is recovered by sucrose gradient centrifugation, detailed in *Protocol 6*. However, partially digested DNA recovered from sucrose gradients is occasionally degraded. This shows as appreciable streaking below 25 kb. Make sure all solutions have been autoclaved (in the case of sucrose, to 110°C). We have never succeeded in constructing a library if there is evidence that some degradation has occurred. We have even seen DNA degrade on 0.2% agarose gels run overnight. This is very hard to control: make sure the gel apparatus is clean and solutions fresh or autoclaved.

An alternative procedure is to dephosphorylate the insert DNA using *Protocol 7*. This is dependent upon a good batch of calf intestinal phosphatase (CIP) being available, and the minimal amount of enzyme being used to produce the desired non-ligatable DNA ends. Our experience has been that CIP not only removes terminal phosphates but appears to also occasionally produce damaged sticky ends. Cloning with phosphatased material is usually less efficient than similar experiments with size fractionated DNA.

Protocol 6. Size fractionation of partially digested DNA

Reagents

- 40% and 10% sucrose (w/v) solutions, in 1 M NaCl, 20 mM Tris–HCl pH 7.1, 20 mM EDTA, 0.3% (v/v) Sarkosyl

Method

1. Pour a 38 ml 40–10% sucrose gradient using a gradient maker approximately 1 h before centrifugation and make sure that there is sufficient

room for the DNA sample: take off the required volume if necessary. (Alternatively, a 'step gradient' comprising equal volumes of 40%, 35%, 30%, 25%, 15%, and 10% sucrose solutions successively layered into the tubes can be employed if left to diffuse overnight before use.) Carefully layer 1 ml of DNA solution on to each gradient, ensuring that there is not more than 300 μg per gradient. We generally use three gradients and one balance. Centrifuge at 26 000 r.p.m. for 16 h at 10°C in a Beckman SW28/27.1 rotor or its equivalent.

2. Fractionate the gradients by piercing the bottom of the tube with a needle attached to tubing through a pump. Collect 800 μl fractions. The DNA of appropriate size should be found in fractions 15–25. Check 15 μl of alternate fractions by electrophoresis on a 0.2% gel. Do not forget to add 40% sucrose solution to the markers tracks, giving a final concentration of approximately 30% sucrose. This is important as the high ionic strength greatly affects DNA mobility.
3. Precipitate DNA from fractions of the correct size by adding 2 vol. of 99% ethanol. To avoid precipitation of sucrose it is a good idea to add a further 2 ml of 70% ethanol. Mix thoroughly and leave at −20°C at least overnight or longer. Pellet the DNA by ultracentrifugation. We routinely use the Beckman SW40.1 rotor at 20 000 r.p.m. for 30 min at 4°C. Carefully wash the pellets twice with 70% ethanol and resuspend in 50 μl of TE. Subject 1 μl of this to electrophoresis on a 0.2% gel to identify the useful fractions (50–40 kb). Add 0.05 vol. 3 M sodium acetate pH 5.5, precipitating the DNA by the addition of 2 vol. of ethanol. Pellet by centrifugation and resuspend the DNA in TE to give a concentration of approximately 1–2 μg/μl.

Protocol 7. Dephosphorylation procedure

Reagents

- Calf intestinal phosphatase (Boehringer Mannheim, 1097075)
- Nitrilo-triacetic acid (NTA) (Sigma, N9877)
- 1 × CIP buffer: 50 mM Tris–HCl pH 9.0, 10 mM EDTA (supplied as a 10 × stock with the enzyme)

Method

1. Carry out the preparation of the partial digests as described in *Protocol 4*, but resuspend the DNA at approximately 200 μg/ml in 1 × CIP buffer after the ethanol precipitation.
2. Add approximately 0.5 U calf intestinal phosphatase (CIP)/μg of DNA, and incubate at 37°C for 20 min. Inactivate the CIP by the addition of 0.1 vol. 0.1 M NTA, followed by incubation at 70°C for 20 min.
3. Immediately precipitate the DNA by the addition of 2 vol. of ethanol,

Protocol 7. *Continued*

followed by incubation on dry ice–ethanol for 20 min. Recover the DNA by centrifugation, and carefully wash the pellet with 70% ethanol.

4. Resuspend the DNA in TE at approximately 1 μg/μl.

2.6 Vector preparation

The enzymes that are used to cleave the cosmid vector in preparation for cloning will depend on the actual cosmid used. In *Figure 1* we show the general procedure for a dual-*cos* cosmid vector. The procedure detailed in *Protocol 8* applies to the Lawrist4 cosmid vector. Other vectors may require the use of different restriction enzymes.

Protocol 8. Preparation of cosmid vector

Reagents

- Uncut cosmid vector (e.g. Lawrist4)
- *Bam*HI (Boehringer Mannheim, 656275)
- *Sca*I (Boehringer Mannheim, 775266)
- Calf intestinal phosphatase (CIP) (Boehringer Mannheim, 1097075)
- T4 DNA ligase (Boehringer Mannheim, 716359)
- Nitrilo-triacetic acid (NTA) (Sigma, N9877)

Method

1. Cut a 10 μg aliquot of Lawrist4 with *Sca*I, using 5 U/μg for 1 h at 37 °C. Use the restriction enzyme manufacturer's conditions for cleavage. After the digestion has been checked for completion by electrophoresis on a 0.8% agarose gel, retain an aliquot (0.2 μg) for the test in step 3.

2. Treat the restriction digest with CIP. Add excess CIP (2 U/μg vector DNA) directly to the restriction digest and continue incubation for 30 min at 37 °C. Inactivate the CIP by the addition of 0.1 vol. 0.1 M NTA, followed by incubation at 70 °C for 20 min. Immediately precipitate the DNA by the addition of 2 vol. of ethanol, followed by incubation on dry ice–ethanol for 20 min. Recover the DNA by centrifugation, and wash the pellet with 70% ethanol prior to resuspension in TE (final DNA concentration of 0.5 μg/ml).

3. Remove a 0.2 μg aliquot of phosphatased vector and check that it will not religate in an overnight ligation, using standard ligation buffers and conditions (*Protocol 9*). The aliquot from step 1 may be used as a control.

4. If no ligation is seen, cut the samples with *Bam*HI. Make sure high quality enzyme is used for the minimum time required for complete digestion. Check for completion of the reaction by the electrophoresis of an aliquot on a 0.8% agarose gel.

5. If the digestion in step 4 is complete, set-up a ligation reaction. Check that ligation is complete by electrophoresis of the reaction mix on a 0.8% agarose gel. A fragment of approximately 8 kb should be seen. There should be nothing of higher molecular weight. We find that ligation buffer and enzyme can cause aberrant mobility of fragments, so run the appropriate controls for this.
6. If all samples ligate correctly, phenol extract the entire reaction mixture from step 4. Precipitate DNA from the aqueous phase with an equal volume of isopropanol, wash with 70% ethanol, and take up in a small volume of TE. Prepared vector will be needed at a concentration of approximately 1 μg/μl.

2.7 Ligation reactions

Ligations should be set-up using the manufacturer's recommended buffers (*Protocol 9*). If the ligation buffer is made up in the laboratory rather than using that supplied with the enzyme, take great care to neutralize the ATP, since the pH of the resultant buffer can be significantly non-optimal if un-neutralized ATP is used.

Protocol 9. Vector–insert ligation

Reagents

- Suitably prepared cosmid vector
- T4 DNA ligase (Boehringer Mannheim, 716359)
- 10 × ligase buffer (supplied with the enzyme)

Method

1. Set-up the following reaction mixture:
   - insert DNA 2.5 μg
   - prepared cosmid vector 1 μg
   - 10 × ligase buffer 2 μl
   - water to 19 μl final volume x μl
   - T4 DNA ligase (1 U/μl) 1 μl
2. Incubate at 22°C for 2 h or 15°C for 17 h.
3. Check an aliquot (0.2 μg DNA) with suitable controls by electrophoresis on a 0.2% agarose gel. Ligation of the vector should be obvious. Ligation of the insert to itself should also be seen, but the latter may not be as obvious since a molar excess of vector has been used.

Protocol 9. *Continued*

4. If ligation of the vector has occurred, proceed to the packaging reaction. If no ligation is seen, the normal problem is carry over of the sucrose and/or Sarkosyl from the gradient. To overcome this, re-precipitate the DNA with 2 vol. of 99% ethanol. After centrifugation, wash the pelleted DNA with 70% ethanol, and air dry. Redissolve in 1 × ligase reaction buffer and repeat the ligation and electrophoresis.

2.8 Packaging

Commercially available packaging extracts have high efficiencies but can be costly. Nevertheless, they are the most convenient, and we have successfully used Amersham and Stratagene GigaPack Gold extracts. Packaging of ligation reactions is described in *Protocol 10*.

Failure of ligation is generally due to carry over of salt, detergent, or sucrose and is easily identified: the vector fails to ligate to itself in the presence of insert DNA. However, failure of packaging is somewhat more difficult to identify. Extracts can be carefully checked for their ability to package λ phage DNA. It is also important to ensure there is no contamination of dilution buffers, etc. with detergent from washing-up or other glassware cleaning fluids. If the packaging extracts are working, and if ligation of the vector is seen, but no colonies are seen after plating out, provided the plating cells are good, the insert DNA must be suspect. In such an event, re-precipitate the insert DNA and repeat the ligation. If this does not work, then the sticky ends of the DNA are damaged and the partially digested DNA will have to be prepared from scratch.

Protocol 10. Packaging of ligation products

Equipment and reagents

- Packaging kit (Stratagene, 200216)
- SM: 10 mM Tris–HCl pH 7.4, 10 mM $MgSO_4$, 0.01% (w/v) gelatin

Method

1. Package 4 μl of the ligation mixture precisely as recommended by the manufacturer.
2. Dilute the extract at the end of the reaction with 1 ml of SM phage dilution buffer.
3. Titre 1 μl and 10 μl of the diluted reaction as described in *Protocol 11*.

2.9 Plating the library

We grow *E. coli* ED8767 to saturation (overnight) at 37°C with vigorous aeration in L broth supplemented with 10 mM $MgSO_4$ and 0.2% (w/v) maltose. This latter supplement increases the expression of MalB, the lambda receptor on the cell surface. Always check that the plating cells are capable of supporting λ phage growth by routinely plating a phage sample on to the cells. It is not uncommon that cells become λ-resistant: these will not support the introduction of cosmids or any other packaged molecule. Always check plating cells for contamination by plating on to L agar containing the selective antibiotic without added extracts. Check all solutions (SM, $MgSO_4$, etc.) in a similar fashion. In particular, be careful of slow growing contaminants (yeast and pseudomonads) which are often indistinguishable from slow growing cosmid-containing *E. coli*. Leaving plates out after analysis for a few days at room temperature can be very instructive in identifying contaminants. The easiest way of introducing such contaminants is by serial passage of plating cells—using one culture to start the next. **Always** go back to the stock. *E. coli* ED8767 is strongly *recA* and rarely reverts, but UV sensitivity should be checked regularly.

Packaging extracts can kill bacteria that carry cosmids. If the proportion of diluted packaging extract to saturated cell culture exceeds 30%, titres of cosmids will be dramatically reduced. We also find that some component of the packaging reaction, probably putrescine, can inhibit growth of bacteria spread on plates, even when present at less than 25%. The following protocol should be scaled up or down as required for analytic or preparative scale platings, but care should be taken not to plate greater than the equivalent of 500 μl of 'completed' packaging reaction per 22 cm × 22 cm plate.

The efficiency of cosmid library construction can be very variable and a final yield of 10^4–10^6 colony-forming units (c.f.u.)/μg of insert DNA can be expected using *Protocol 11*. We generally achieve between 5×10^4 c.f.u. and 5×10^5 c.f.u. To put this in perspective, a complete human library, as defined by the statistics of Clarke and Carbon (29), requires 344 000 cosmids for a 99% representation (assuming a vector capacity of 40 kb and a genome size of 3×10^9 bases). Even in unamplified libraries, under-representation and over-representation of individual clones has been seen. The basis of this is not entirely clear (21, 22) but it can effectively render a sequence 'unclonable'. Underrepresentation by even a factor of two decreases the probability of a clone being contained within an otherwise representative library. One source of variability, namely transcriptional interference of the origin by promoter homologies in the inserted DNA, can be partially reduced by using cloning sites protected by RNA polymerase terminators (21).

Protocol 11. Adsorption and plating of packaged cosmids

Reagents

- Overnight culture of *E. coli* ED8767 grown in 10 mM $MgSO_4$/0.2% maltose

Method

1. For every 1 ml of packaging reaction, add 10 ml of a fresh saturated culture of *E. coli* ED8767. Leave at 37°C for 30 min.

Protocol 11. *Continued*

2. Dilute with 25 ml of pre-warmed (37°C) L broth (containing no selective antibiotic) and incubate at 37°C for 1 h.
3. Centrifuge the cells at low speed (5 min at 2000 r.p.m.) in a preparative centrifuge such as a Sorvall RC-5 or its equivalent and resuspend in 1.2 ml of L broth.
4. Plate no more than 500 μl of this mix per 22 cm × 22 cm plate.

2.10 Choice of host genotype

There are a number of reports that recombination-deficient or mutated strains of *E. coli* can affect the stability of DNAs cloned into a variety of vectors. Practically, the instability presents as loss of insert, with a number of deletion derivatives often present (12). (We have observed this phenomenon in the presence and absence of selection, though in some cases, the degree of instability appears to be lower if suspect clones are grown on plates rather than in liquid culture.) The use of mutated strains should be approached with caution as several genotypes cannot support cosmid growth. Below we list the strains that we have shown can support cosmid growth to varying degrees of efficiency, though some are unlikely to provide useful host systems as growth rates are severely retarded. Our personal view is that no great reliance should be placed upon strains other than *recA E. coli*. The best-documented strain effects are for the *recBCsbcBsbcC* mutations (30) that are reported to stabilize head-to-head repeats. In our hands, strains of this genotype will not support cosmid growth. In every other case, strain stabilization has been anecdotally reported. To our knowledge, no physical basis has been described for the underlying instability of some cosmid clones, and therefore the information remains anecdotal. However, in desperation, it would be worth selecting any of the genotypes listed in *Table 2* in an attempt to stabilize an unstable cosmid, provided a limited expectation of success was acceptable! This area remains a baffling and frustrating area of our understanding of prokaryotic cloning techniques.

As an additional cautionary tale, human DNA cloned in yeast artificial chromosomes has also been observed to show instability. The sequences that exhibited instability were tandem repeats (31), unfortunately a relatively common characteristic of eukaryotic genomes. Recent studies of tracts of simple repetitive DNA in yeast have indicated that instability is increased by mutations in genes responsible for mismatch repair, while mutations that affected the proof-reading ability of the DNA polymerases had little or no effect (32). In *E. coli*, mutations affecting recombination (*recA*) appear to have no effect on tract stability, while *E. coli* strains with *mutL* and *mutS* genotypes (mismatch repair) show considerable tract instability. Just exactly how these factors contribute to the overall instability of some genomic DNA

Table 2. *E. coli* genotypes that may be used in an attempt to stabilize inserted DNAs in cosmid vectors

Genotype	Growth
recA	+
recBCsbcBsbcC	−
recBCsbcBsbcCrecA	−
recBCsbcBsbcCrecF	−
recBsbcBsbcC	−
recCsbcBsbcC	slow
sbcC	slow
sbcCrecA	+
recD	+
recF	+

sequences cloned in *E. coli* remains unclear. We have, until now, proceeded upon the not unreasonable view that determinants of instability in prokaryotic and eukaryotic cells would be different. It thus seems likely that for any cosmid mapping project, there is likely to be a small proportion of the region of interest that is refractive to cloning.

2.11 Library manipulation

2.11.1 Amplification

Generally, it is possible to spread 500 μl of the cells (prepared as detailed in *Protocol 11*) on a 22 cm × 22 cm plate. This should give, from an average experiment, approximately 50 000 colonies per plate. Pour thick (300 ml/plate) L agar plates supplemented with the appropriate drug. If possible, plate the library on to plates that are one or two days old, as they will absorb more liquid, and aid the spreading process. After spreading, allow the plates to dry, if necessary, in a sterile tissue culture hood and then incubate them upside-down for 24 h at 37°C.

Colonies may be scraped from the plate by adding 2 ml of L broth plus 15% glycerol before scraping. Mix the pooled colonies carefully and store at −70°C or in liquid nitrogen. Titres of bacteria should be approximately 1–2 × 10^9/ml, thus generating many 'libraries' worth'.

Amplified libraries sometimes do not show random representation of clones. Under and over-representation of clones occurs but this may not be a primary concern. Some bacteria containing cosmids will always grow to give small colonies. Other cosmid DNAs are unstable, undergoing spontaneous rearrangement or deletion of sequences to become smaller. Typically, this occurs in approximately 2–5% of recombinants. The cause of this phenomenon is difficult to pin down, and may be any one of several reasons. Inverted repeats, strong prokaryotic promoter homologies in the eukaryotic DNA, operator homologies, and repressor binding sites are all candidates (see ref. 21 and above for a discussion of these phenomena). We have identified cosmids that cause their host cells to grow slowly on plates and in bulk culture. Similarly, deletion derivatives would have a growth advantage, leading to a mixed population of variants. In practical terms, there is little that can be done about extremely unstable cosmids except to attempt to recover and store full-length molecules from the deletion culture.

This can be achieved by using *in vivo* packaging by superinfection, either on plates or in solution (*Protocol 19*). Cosmids recovered from the phage lysates will be enriched for full-length cosmids, since they will fall within the correct packaging limits. However, multimers of deletion derivatives can also be packaged, and as such this process may not be efficient. To minimize deletion events/clones, it is important to store cosmids as soon as possible after isolation (e.g. store some of the bacterial suspension used for the initial mini-preps) and grow cosmid stocks as little as possible. Storing packaged, but untransformed, cosmids is another alternative, although titres do not remain high.

2.11.2 Plating on to filters

Cosmid libraries can also be plated directly on to filters. However, all the filters we have used, both nylon and cellulose nitrate, are mildly toxic to bacteria. Washing the filter in water and sterilization helps but does not eliminate the problem. If you are planning to plate the library on filters, always titre the library on filters: do not be tempted to extrapolate directly from the results obtained by plating directly on to selective agar.

We have obtained the best results with cosmids that have been absorbed to *E. coli* and then plated exactly as described in *Protocol 11*. A reduction in titre of 20% can be expected and in some cases may be as much as 90%. Some *E. coli* strains do not grow at all well on nylon membranes. It is strongly advised that the *E. coli* host cells be tested for their ability to grow both on filters and directly on L agar containing antibiotic.

Perversely, the scale up of packaging/ligation reactions is often not a matter of simple multiplication: tenfold more extract does not always give tenfold more cosmids. Be very careful to control scale up—scale up *exactly* as the conditions of the small scale reaction dictate. Despite these points, *E. coli* ED8767 appears to be quite a reliable host. If the library is plated directly, then it is important to follow *Protocol 12* for the optimal preparation of replicas. The time allowed for the growth of colonies is particularly critical, since we find that bacteria on filters tend to grow slower than on agar.

2.11.3 Replica plating for library screening

The reader should refer to Chapter 2 for a detailed consideration of how to grid out cosmid libraries for screening. Here we describe an approach for screening cosmid clones that are randomly distributed on agar plates as a result of the plating process.

We prefer to work with nylon Pall/Biodyne 1.2 micron membrane filters and use the methods of Hanahan and Meselson (33), suitably modified, to make master and replica filters. We have found it very difficult to mix filter types in making replicas (e.g. Biodyne does not replicate well on to nitrocellulose). Therefore, we use only a single filter type for all operations. Additionally, the Pall membranes appear to retain bound DNA well, enabling repeated hybridizations of the library without fear of DNA loss. The instructions in *Protocol 12*, which draw heavily on those of Bernard Herrman (Tubingen) and Thomas Pohl (Jackson Laboratories), refer to the use of an amplified library stock but need only appropriate and obvious modifications for primary libraries.

Protocol 12. Preparation of replica filters for storage and hybridization [a]

Equipment and reagents

- Three 22 cm × 22 cm L agar plates: L agar is 10 g Difco bacto tryptone, 10 g Difco yeast extract, 5 g NaCl, 15 g agar made up to 1 litre with water—autoclave and add appropriate antibiotic when the solution has cooled to 50°C immediately before pouring
- Three 22 cm × 22 cm F agar plates: F agar is L agar plus 5% glycerol—autoclave, add the appropriate antibiotic, and pour as above
- Lysis solution: 0.5 M NaOH, 1.5 M NaCl
- Neutralizing solution: 3 M sodium acetate pH 5.5
- Two 4 mm Perspex sheets (25 cm × 25 cm)

A. *Preparation of replica filters*

1. Make dilutions of the amplified stock. 3.5 ml of broth containing approximately 1×10^6 cells/ml will be required. Titre this by spreading on to filters placed on L agar plates containing the selective antibiotic.
2. The next day, place a filter, with a number written in pencil at one side, on to a fresh F agar plate containing antibiotic and allow the filter to become wet. (Do not use plates that are very wet: dry them first to remove obvious excess moisture.) Spread sufficient bacteria to give 10^6 colonies per plate (3×10^6 in total), using the **same** dilutions as in step 1. Make sure you leave an approximately 0.5–1.0 cm gap between the spread bacteria and the edge of the filter, taking care to maintain sterility. Check that the liquid is being absorbed into the filter. When dry, incubate the plates upside-down at 37°C for approximately 12 h. (It is possible to incubate the plates at 30°C to slow the growth: typically 16–18 h at 30°C.)
3. After 12 h, the colonies should be small (approximately 0.5 mm). Size is critical, and thus the incubation time should be adjusted as necessary.
4. Place the required number of filters, labelled in pencil, on to L agar plates containing antibiotic to pre-wet.

Protocol 12. *Continued*

5. Put some filter paper (Whatman 3MM) on to a flat surface and place the master filter, colonies up, on to this.
6. Very carefully put a second filter on top of the master: do not adjust the position if they are not perfectly superimposed.
7. Put a second sheet of 3MM paper on top of the filters and a strong glass plate on top of this. Press down hard and uniformly over the plate.
8. Remove the plate and the top 3MM paper, and with a needle carefully stab the two filters to make alignment marks. Make two stabs close together and seven to ten others asymmetrically over the filters.
9. Peel apart the two filters with a single reasonably firm motion and place the filters, colonies up, on to two plates: the master back on the F agar plate, and the replica on the L agar plate.

B. *Preparation of the master filter for long-term storage*

1. The master filter can now be frozen away for further use or stored at 4°C for up to one month. For freezing at −70°C, the master filter should be incubated at 37°C until the colonies are beginning to show signs of growth. Place the filter on a sheet of Whatman 3MM that has been soaked in L broth containing 15% glycerol. After 10 min, place the filter on an acetate sheet the same size as the filter. (We typically use overhead acetate sheets for this.) Photocopy a standard 1 mm grid on to another acetate sheet and place the gridded acetate carefully on to the colony filter. Do not move the acetate grid once it has adhered to the colonies. Gently remove any air bubbles that may have formed by applying an even pressure over the acetate grid.
2. Transfer all the positional alignment marks on the colony filter to the gridded acetate, indicating the alignment mark holes as crosses on the acetate with a fine-tipped waterproof marker pen. Now place another gridded acetate on top of the first one and repeat the alignment marking procedure. As this acetate will be used to identify the position of positive clones, ensure that it matches the primary grid exactly. Make as many copies as you feel necessary.
3. Place the acetate/filter/gridded acetate sandwich, along with two 25 cm square 3 mm Perspex sheets, at −20°C for 30 min. Assemble the stack, placing the filter sandwich between the Perspex sheets, and clamp together with eight small Bulldog clips. Store at −70°C, preferably in a polythene bag to minimize the chance of the assembly being disturbed inadvertently.

C. *Preparation of replica filters for hybridization*

1. Replica filters should be grown for approximately 3–5 h at 37°C. The colonies will again be approximately 1 mm diameter or less. These can

then be processed for screening. Colonies are lysed and processed exactly as described in the Pall/Biodyne literature. Place the filter colony side up on to Whatman 3MM paper soaked in lysis solution for 10 min, followed by 10 min on 3MM soaked in neutralizing solution. Air dry and bake at 80°C for 1 h. UV cross-linking is probably not necessary, but can be performed if so wished. Store the dry filters sealed in a polythene bag at room temperature.

2. Rinse the filters in an excess of 5 × SSC prior to pre-hybridization.

[a] The method detailed is based on that of Bernard Herrman and Thomas Pohl (personal communication).

2.11.4 Screening cosmid filters

Filters can be screened by a variety of methods. We favour one which allows use of either RNA or DNA probes without significant changes to the technique (*Protocol 13*). Probes must be of high specific activity, > 10^8 c.p.m./μg, although the method of labelling is unimportant. We routinely use random primed labelling for DNA probes (34). Positive clones can be picked into microtitre plates, an efficient method of storage, and screened using gridding technologies (26, 35), methods that are described in detail in Chapter 2.

Protocol 13. Hybridization of filters

Reagents

- 100× Denhardt's solution: 2% (w/v) Ficoll 400, 2% (w/v) bovine serum albumin, 2% (w/v) polyvinylpyrrolidine
- 1 M sodium phosphate buffer pH 6.5
- 10% SDS
- Sonicated salmon sperm DNA (10 mg/ml) (Sigma, D9156)
- Deionized formamide
- Poly-d(C) (Boehringer Mannheim, 108715)
- Poly-d(A) (Boehringer Mannheim, 223581)
- *Sau*3AI digested cosmid vector
- ^{32}P-labelled probe of interest (DNA or RNA)
- Coomassie dye mix: 1 g Coomassie brilliant blue, 500 ml isopropanol, 500 ml glacial acetic acid, 1.3 litres water
- Vanadyl ribonucleoside complexes (VRC) (Sigma, R3380)

Method

1. Wash the filters in an excess of 5 × SSC for 5 min at room temperature.
2. Pre-hybridize in 20 ml of the following solution (HybeMix):[a]
   - 5 × Denhardt's solution
   - 5 × SSC
   - 50 mM sodium phosphate buffer pH 6.5
   - 0.1% SDS
   - 250 μg/ml salmon sperm DNA, denatured and sonicated
   - 10 μg/ml poly(C)
   - 10 μg/ml poly(A)
   - 50% formamide (deionized)

Protocol 13. *Continued*

3. For DNA hybridizations, pre-hybridize at 42°C for at least 4 h. For RNA hybridizations, a 1 h pre-hybridization is usually sufficient.
4. Hybridization is carried out in 15 ml of HybeMix. The buffer has the same composition as step 2 save that it contains 10^6 c.p.m./ml of the appropriate ^{32}P-labelled DNA or RNA probe. For RNA probes, supplement the reaction with 10 mM VRC and hybridize for 17 h at 50°C. For DNA, hybridize at 42°C for 48 h.
5. After hybridization, filters are washed identically, independent of whether DNA or RNA probes are used. Initially, three washes of 20 min each in 500 ml of 2 × SSC, 0.1% SDS are performed to remove unbound probe. The final stringent wash is typically with 500 ml 0.1 × SSC, 0.1% SDS for 30 min at 65°C. This stringent wash is for homologous probes and can be modified as required to suit the experiment.
6. Expose the filters to Kodak XAR5 (or Fuji RX) film with intensifying screens at −70°C for one to four days.
7. Identification of positively hybridizing clones can be facilitated by staining colony debris on the filter (36). Place the filter in a large Petri dish and add sufficient Coomassie dye mix to cover the filter. Agitate the filter gently for 5 min, and then rinse extensively in water. Air dry the filter: the background will fade.
8. Photocopy the colony pattern on to overhead transparencies, and mark the position of the key marks. Align the autoradiograph with the photocopied transparency, and mark on the transparency the positions of the positive clones.
9. Pick the required clones into L broth containing the suitable antibiotic.

[a] It is useful to include denatured restriction digested (e.g. *Sau*3AI) cosmid vector as competitor in the pre-hybridization at 1.0 μg/ml.

2.11.5 Storing cosmid libraries

Cosmids can be stored as packaged phage particles for at least six months at 4°C, although this is not advised: drastic reductions in titres occur after even short periods of time. Plated cosmids can also be stored at 4°C for extended time periods, with minimal loss of viability. In this instance, the growth of contaminating fungi can be a problem.

Amplified libraries are stable virtually indefinitely when frozen at −70°C or in liquid nitrogen in L broth containing 15% glycerol. Aliquot the amplified library (see above) into 1 ml vials and store at −70°C. Thaw out as required. We have seen no obvious reduction in titre after three rounds of thawing of an aliquot, but do not recommend excessive freeze–thawing. The best method of storage is to plate the library directly on to filters, and make −70°C freezer stocks as described in *Protocol 12.*

2.12 Cosmid DNA preparation

There are numerous protocols for preparing plasmid DNA from bacterial cells. The one we describe here is simple and reasonably reproducible but is almost certainly replaceable by other techniques (e.g. column methods). Cells should be cultured in L broth containing the appropriate antibiotic with vigorous aeration. Never grow cosmid-containing cells in the absence of antibiotic. For 'mini'-preps grow 5 ml in 25 ml tubes, while for 'maxi'-preps, use 500–1000 ml in a 2 litre flask shaken vigorously. Start the growth by inoculation of a single colony and incubate overnight at 37°C. Some cosmids grow badly and may not be saturated within this period, in which case incubate for longer. We do not recommend chloramphenicol amplification for non-λ origin cosmids. This treatment may increase yields but also may select for deletion derivatives of the parent recombinant.

With care, the mini-prep (*Protocol 14*) will yield DNA that can be reproducibly cleaved with restriction endonucleases. If you find that this is not the case, phenol extract the preparation, ethanol precipitate the nucleic acids, wash the pellet with 70% ethanol, redissolve it, and try again. This is not routinely necessary. The most common causes of mini-prep failure (partial restriction digestions or complete degradation) are the incomplete removal of precipitated chromosomal DNA or co-precipitated salt.

Protocol 14. Cosmid DNA mini-preps

Reagents

- Solution I: 25% (w/v) glucose, 50 mM Tris–HCl pH 8.0, 10 mM EDTA
- Solution II: 0.2 M NaOH, 1% (w/v) SDS—make this fresh and use NaOH solutions that have been stored in plastic bottles, not glass
- Solution III: 3 M potassium acetate pH 4.8 with acetic acid
- 5 M LiCl

Method

1. Pellet the cells from 1.5 ml of saturated culture by centrifugation in a microcentrifuge for 1 min at room temperature.
2. Pour off the culture medium and swab out any excess with a cotton bud. Add 200 μl of cold solution I to the pellet. Resuspend the cells carefully. Leave the tube(s) for 5 min at room temperature.
3. Add 400 μl of freshly made solution II at room temperature. Mix by inverting three times. Do not vortex. The solution will clear as the cells lyse. Place the tube(s) on ice for 10 min.
4. Add 200 μl of cold solution III, mix by shaking, but do not vortex. A

Protocol 14. *Continued*

large white precipitate should form immediately. Leave the tube(s) on ice for 10 min.

5. Centrifuge in a microcentrifuge at room temperature for 10 min.
6. Pick out the rather sticky and gelatinous pellet with a toothpick. The pellet is often smeared over the side of the tube and looks horrible—don't worry, this is normal.
7. Add 480 μl of propan-2-ol at room temperature and mix well. Freeze on dry ice and then transfer whilst still frozen to a microcentrifuge. Centrifuge for 10 min at room temperature.
8. Pour off the supernatant—there should be a clearly visible pellet. Leave the tubes upside-down to drain for a couple of minutes. Swab off the excess isopropanol with a cotton bud. Resuspend the pellet thoroughly in 100 μl of water by incubating the tube(s) at 65°C for 10 min.
9. Add 100 μl of ice-cold 5 M LiCl, and mix quickly. Incubate on ice for at least 10 min, during which time the precipitating RNA and traces of debris may be seen.
10. Centrifuge the tube(s) in a microcentrifuge for 10 min, and remove the supernatant to a fresh tube. Add 2 vol. of ethanol, and precipitate the DNA at −20°C for at least 10 min.
11. Centrifuge the tube(s) in a microcentrifuge for 10 min, and discard the supernatant. Fill the tube(s) with 70% ethanol, vortex briefly to dislodge the pellet, and re-centrifuge the tube(s) for 5 min.
12. Discard the supernatant, being careful to ensure that the pellet is not lost. Excess 70% ethanol can be removed using a Gilson Pipetman. Allow to air dry for 5–10 min.
13. Resuspend the pellet in 50 μl TE.
14. For restriction enzyme analysis of mini-prep DNA, digest 5 μl of the DNA in a final volume of 20 μl.

Yields from the maxi-prep (*Protocol 15*) are often disappointing: much less than that obtained from mini-preps would lead you to believe. However, 500 ml of culture should yield 300–700 μg of cosmid, depending on the vector used. Yields of non-recombinant λ origin cosmid vectors are also low: we rarely get more than 200 μg/litre. This is due to the nature of replication control of λ derivatives. While the lower yield is an annoying facet of working with some cosmid systems, it does have the concomitant bonus of a lower incidence of unwanted recombination events: the cosmid is generally more stable.

Protocol 15. Cosmid DNA maxi-preps

Reagents

- Solution I: 25% (w/v) glucose, 50 mM Tris–HCl pH 8.0, 10 mM EDTA
- Solution II: 0.2 M NaOH, 1% (w/v) SDS—make this fresh and use NaOH solutions that have been stored in plastic bottles, not glass
- Solution III: 3 M potassium acetate pH 4.8 with acetic acid
- CsCl
- Water-saturated butan-1-ol

Method

1. Pellet the cells from a 500 ml culture by centrifugation and add 5 ml of cold solution I to the pellet. Resuspend thoroughly using an inoculation loop. Add a further 15 ml of solution I and a small amount of lysozyme crystals (the tip of a spatula's worth). Mix and leave at room temperature for 10 min.
2. Add 60 ml of solution II, and mix by inverting a couple of times.
3. Add 45 ml of cold solution III. Mix as in step 2. Leave on ice for 30 min.
4. Centrifuge for 15 min at 5000 r.p.m. in a GSA rotor of the Sorvall RC-5 (or equivalent). Transfer the supernatant to a fresh tube, removing clumps of debris by pouring through gauze. Add 0.6 vol. of propan-2-ol, mix, and leave on ice for 30 min. Clouds of precipitating nucleic acids will be seen. Recover the precipitate by centrifugation as above. Drain off the excess propan-2-ol and resuspend in TE. (The volume of TE can be modified to take into account the size of ultracentrifugation tube that will be used for the CsCl gradient. We typically use the Beckman VTi65 rotor, and thus resuspend the pellet in 5 ml TE, dividing the sample between two tubes for ultracentrifugation.) Measure the volume of the dissolved nucleic acids.
5. If the volume of DNA solution is x ml, add 0.1 x ml of 10 mg/ml ethidium bromide, followed by 1.1 x g CsCl. Ensure that the CsCl has dissolved: it is extremely endothermic. Incubate in the dark at room temperature for 30 min, during which time a heavy precipitate will form. Centrifuge the sample for 10 min at 5000 r.p.m. in a SS34 rotor of the Sorvall RC-5 (or equivalent), ensuring that the temperature of the centrifuge does not fall below approximately 14°C: if it does, the CsCl will precipitate.
6. Carefully recover the supernatant, ensuring that little or no debris is retained. Apportion the sample into two ultracentrifuge tubes, ensuring that they are balanced to within 0.1 g. Top up the tubes, excluding all air, with liquid paraffin. Seal the tubes by crimping/heat sealing. We typically centrifuge the samples for 17 h at 50 000 r.p.m. at 20°C in a Beckman VTi65 rotor.

Protocol 15. *Continued*

7. After centrifugation, three intensely fluorescing regions or bands will be seen in the gradient. The uppermost represents *E. coli* chromosomal DNA and linear and relaxed-circle plasmid, the second band represents supercoiled plasmid, while the lowest 'band', which should have pelleted on the bottom of the tube, is the RNA. Harvest the second (middle) band using a syringe and needle, ensuring that the top of the tube has been punctured prior to harvesting. This prevents the formation of a vacuum within the tube, which leads to bubbles of air dislodging the gradient, as the plasmid DNA is harvested. Harvest the supercoiled DNA in the smallest volume possible, putting the solution into a 50 ml polypropylene tube (e.g. Falcon 2070). Measure the volume as *x* ml.
8. Fill the tube with water-saturated butan-1-ol, cap, and shake. Allow the phases to separate, or alternatively, centrifuge the tube in a bench-top centrifuge for a few minutes at a medium speed. Discard the upper (now pink) organic phase into a suitable container, and repeat this process until the lower aqueous phase is completely clear. Remove this layer into a fresh tube and measure its volume. Add sufficient distilled water to give a volume of 4 *x* ml.
9. Add **either** 2 vol. of ethanol **or** an equal volume of propan-2-ol, mix, and incubate at −20°C for 1 h. Recover the precipitated plasmid DNA by centrifugation for 10 min at 10 000 r.p.m. at 4°C in a SS34 rotor in a Sorvall RC-5 (or its equivalent).
10. Wash the pellet carefully in 70% ethanol at room temperature, air dry for a few minutes, and resuspend the DNA in 500 μl of TE by incubating at 65°C for 10 min. Measure the concentration by spectroscopy at 260 nm or by agarose gel electrophoresis with known standards.

2.13 Walking

Vectors that contain SP6, T7, or T3 promoters immediately adjacent to the cloning sites enable the simple preparation of ^{32}P-labelled RNA probes that are specific for the boundaries of the cloned DNA. These can be used to identify new cosmids that overlap the starting cosmid. This process of isolating overlapping clones is correctly known as 'walking'. The use of the SP6/T7/T3 promoter systems can significantly reduce the time and labour involved: there is no need to identify ends of clones by restriction mapping prior to probe preparation. The transcription reactions are most efficiently carried out using a CsCl purified template (see *Protocol 16*).

RNA transcripts that are specific for the ends of the cloned DNA can be made either by cleavage with a restriction enzyme or by the use of limiting UTP concentrations. At 0.5 μM UTP, the transcripts average approximately

800–1000 bp without a run-off cleavage site. RNA probes are very efficient for screening libraries: they tend to give higher signal-to-noise ratios than double-stranded radiolabelled DNA probes but are, of course, more sensitive to degradation. The hybridization conditions employed are detailed in *Protocol 13*.

Protocol 16. *In vitro* transcription of cosmids

Reagents

- Water-saturated butan-1-ol
- [^{32}P]UTP (Amersham International, PB20383)
- Spermidine–HCl
- Dithiothreitol
- Bovine serum albumin (Sigma, B2518)
- ATP (Boehringer Mannheim, 1140965)
- GTP (Boehringer Mannheim, 1140957)
- CTP (Boehringer Mannheim, 1140922)
- SP6 RNA polymerase (Boehringer Mannheim, 1487671)
- T3 RNA polymerase (Boehringer Mannheim, 1031171)
- T7 RNA polymerase (Boehringer Mannheim, 881775)
- 5 × transcription buffer: 200 mM Tris–HCl, 30 mM $MgCl_2$, 10 mM spermidine–HCl, 500 μg/ml bovine serum albumin pH 7.5
- Vanadyl ribonucleoside complexes (VRC) (Sigma, R3380)
- RNasin (Promega, N2511)

Method

1. Recover the plasmid band from the CsCl gradient in *x* ml.
2. Extract twice with 50 ml of water-saturated butan-1-ol as detailed in *Protocol 15*. Measure the volume of the DNA solution as *x* ml.
3. Add *y* ml of water to give a final volume of *z* ml (four times *x*).
4. Add an equal volume of propan-2-ol, mix well, freeze on dry ice, and centrifuge to pellet the DNA.
5. Wash the pellet with 70% ethanol at room temperature.
6. Collect the pellet by centrifugation, allow it to air dry, and resuspend in 500 μl TE.
7. Use 20 μg of recombinant cosmid per transcription reaction (equivalent to approximately 2 μg of a small pUC-derived vector). If a specific run-off site is required, cut with the appropriate enzyme, extract the DNA once with phenol, and once with chloroform. Precipitate the DNA with propan-2-ol, wash with 70% ethanol, dry, and resuspend in a suitable volume of TE. If cleavage is not required, the DNA can be used directly.
8. Transcription reactions should be set-up at room temperature to avoid spermidine precipitation. Set-up 100 μl reactions with the following components (final concentrations/amounts; some of the reagents can be kept as a 5 × concentrated buffer, sterilized by filtration, and stored at 4°C):
   - 20 μg DNA
   - 40 mM Tris–HCl pH 7.5

Protocol 16. *Continued*

- 6 mM $MgCl_2$
- 2 mM spermidine–HCl
- 100 μg/ml bovine serum albumin
- 10 mM dithiothreitol
- 100 U RNasin
- 400 μM ATP
- 400 μM CTP
- 400 μM GTP
- 50 μCi [^{32}P]UTP (800 Ci/mmol) (cold UTP as required)
- 40 U SP6, T3, or T7 polymerase

9. Take an aliquot for assay before the addition of enzyme. Incubate at 40°C for 60 min, and take another aliquot.

10. It is important to assay the transcription reactions since low efficiency of incorporation normally signifies degraded probes that are not suitable for hybridization. Assay by spotting aliquots on to Whatman DE81 filters, and precipitating nucleic acids with 10% trichloroacetic acid. Wash filters with ethanol and count. Incorporation in the absence of cold UTP can be up to 90%.

11. Meanwhile, terminate the reaction by adding an equal volume of chloroform. Shake and separate the phases by centrifugation. Precipitate nucleic acids from the aqueous phase by the addition of an equal volume of propan-2-ol. Pellet RNA by centrifugation and resuspend in 20 μl TE containing 10 mM vanadyl ribonucleoside complexes (VRC) or RNasin (1 U/μl). Use immediately, if possible. Do not denature or attempt to remove the template—it is not necessary.

2.14 Gel electrophoresis

Gel electrophoresis is used at a number of stages in the preceding protocols to analyse high molecular weight DNA molecules. In this section, we present protocols for conventional (*Protocol 17*) and field inversion gel electrophoresis (FIGE; *Protocol 18*).

Protocol 17. 0.2% agarose gels

Reagents

- High gelling temperature agarose (IBI, IB70042)
- 10 × TBE: 106 g Tris base, 55 g boric acid, 9.5 g disodium EDTA per litre pH 9.3
- Ethidium bromide

Method

1. Gels can be made using any standard agarose gel tank. Gels should preferably be cast in the bespoke gel former since they are quite difficult to manipulate once the 0.2% agarose has set.
2. Use a polycarbonate box, for example, that fits inside the normal gel area, leaving a 2 cm gap all around.
3. Place a weight on the box and pour 2% agarose in 1 × TBE to a 0.75 cm depth around it. This will act as a rigid support for the 0.2% gel. There is no need to cover the bottom of the well with 2% agarose.
4. When the agarose has solidified, carefully remove the box. Place the slot former in position approximately 2 cm away from the supporting agarose, pushing the teeth through the edges of the 2% agarose. Fill the central well area with 0.2% agarose in 1 × TBE and leave to solidify at 4°C.
5. Once the gel has set, remove the slot former very gently—the slots will collapse, but open up once submerged under electrophoresis buffer.
6. Very gently flood the electrophoresis container with 1 × TBE electrophoresis buffer—do not pour buffer directly on top of the 0.2% agarose as it will destroy the gel. It is better to include ethidium bromide (0.5 μg/ml) in the running buffer, which is considerably easier than trying to stain the gel after the run. It will, however, reduce the resolution of the gel very slightly.
7. The gels are very sensitive to overloading and a few practice runs will establish the correct volume and amount of DNA to be loaded. Use slots that are 0.75 × 0.15 cm embedded in a gel of 0.75 cm depth. These will hold approximately 25–35 μl comfortably and will fractionate approximately 100 ng of DNA without overloading. Be careful, since an overloaded gel can show DNA running with both greater and lesser mobility than expected.
8. Rapid electrophoresis can be used to determine approximate sizes of DNA fragments. However, for optimal resolution, use a low voltage (1 V/cm). Markers are most conveniently bacteriophage λ DNA run either as multimers or digested to give a 25–30 kb fragment (e.g. λ cleaved with *Kpn*I). The λ multimers do not resolve particularly well on the 0.2% gels described here but clear resolution of 48 kb and 96 kb should be possible.

Protocol 18. Field inversion gel electrophoresis (FIGE)

Reagents

- See *Protocol 17*

Method

1. Pour a 0.8% agarose gel in 0.5 × TBE. The lower TBE concentration is required to minimize the heating effects of the current.
2. Ensure that the 0.5 × TBE buffer in the electrophoresis tank has been cooled to 10°C: a circulation pump, connected to a cooling unit is required for the entire duration of the electrophoresis run.
3. Place the gel in the tank; load samples and suitable size markers (detailed above).
4. Electrophorese the samples for 16 h at 180 V, with a pulse ratio of one-third (reverse to forward). For maximal separation of the 25 kb–250 kb size range, we have found that a continually cycling ramp from 0.15 sec to 12.03 sec is suitable.
5. Stain the gel in ethidium bromide (5 μg/ml) for 15 min, and destain in distilled water for 20 min.

3. Specialist techniques

This section covers in broad outline techniques that may prove useful to some workers in the field. It is not intended to be a detailed protocol for all aspects of the techniques.

3.1 Superinfection

Unstable cosmids are difficult to rescue since both deleted and full-length molecules confer drug resistance. Physical methods are not particularly satisfactory, particularly since cosmids can multimerize within the cell giving an illusion of size. Superinfection with a lambda phage, resulting in *in vivo* packaging then can select full-length molecules from complex mixtures (*Protocol 19*). Success cannot be guaranteed since, again, multimers of deleted cosmid can be packaged.

Superinfection is carried out by infecting a growing culture with a lytic lambda phage; vector phages typically employed for general cloning procedures in most laboratories can be used for this purpose. The choice of the superinfecting phage is arbitrary when non-lambda origin cosmids are used. The Lorist cosmids, however, render the *E. coli* host immune to λ phages of the λ immunity group. Superinfection should be carried out with an imm^{434}

or *imm*[22] immunity group phage. Many vector phages have these immunity groups: see ref. 17 for a suitable choice.

Protocol 19. Packaging by superinfection

Reagents

- 100 ml L broth
- Top agarose: L broth plus 0.7% agarose
- SM buffer: 10 mM Tris–HCl pH 7.4, 10 mM $MgSO_4$, 0.01% (w/v) gelatin
- Superinfecting phage

A. *Liquid cultures*

1. Inoculate 100 ml of L broth supplemented with the required antibiotic and 10 mM $MgSO_4$ with the clone(s) of interest. (It is possible to use the entire library for superinfection.) Inoculate with sufficient bacteria to give an initial A_{600} of approximately 0.08. Grow with vigorous aeration at 37 °C until the A_{600} is 0.4 (2–5 h).
2. Infect the culture with any λ phage that is capable of lysogeny (i.e. cI^-). We have used several phage vectors with equal efficiency.
3. Monitor cell growth at A_{600}. It will increase to 1.0, and then collapse to approximately 0.5 as lysis occurs, usually within 3 h. When lysis occurs, add 1 ml of chloroform, shake for 2 min, and remove the bacterial debris by centrifugation at 4000 r.p.m. in a standard bench-top centrifuge.
4. Titre the supernatant as described in *Protocol 11*.

B. *Plate cultures*

1. Plate 200 μl of a saturated culture of the clone of interest in 2.5 ml top agar on to a standard 90 mm L agar plate containing the required antibiotic.
2. After the top agar has solidified, spot 5–10 μl (10^4–10^5 p.f.u.) of an appropriate superinfecting phage in SM buffer (see above) on to the centre of the plate. Incubate the plate at 37 °C overnight.
3. Pick the clear plaque that has formed, using the wide end of a Pasteur pipette, and place the plug in 1 ml of SM buffer. Add three drops of chloroform, shake gently, and leave at 4 °C for 2–5 h. Titre the supernatant as described in *Protocol 11*, ensuring that chloroform droplets are not removed: they will kill the plating cells.

3.2 Cosmid restriction enzyme site mapping

Cos mapping was developed by Rackwitz *et al.* (37) to map phage λ clones and has been extended to cosmid clones (19). The major problem with

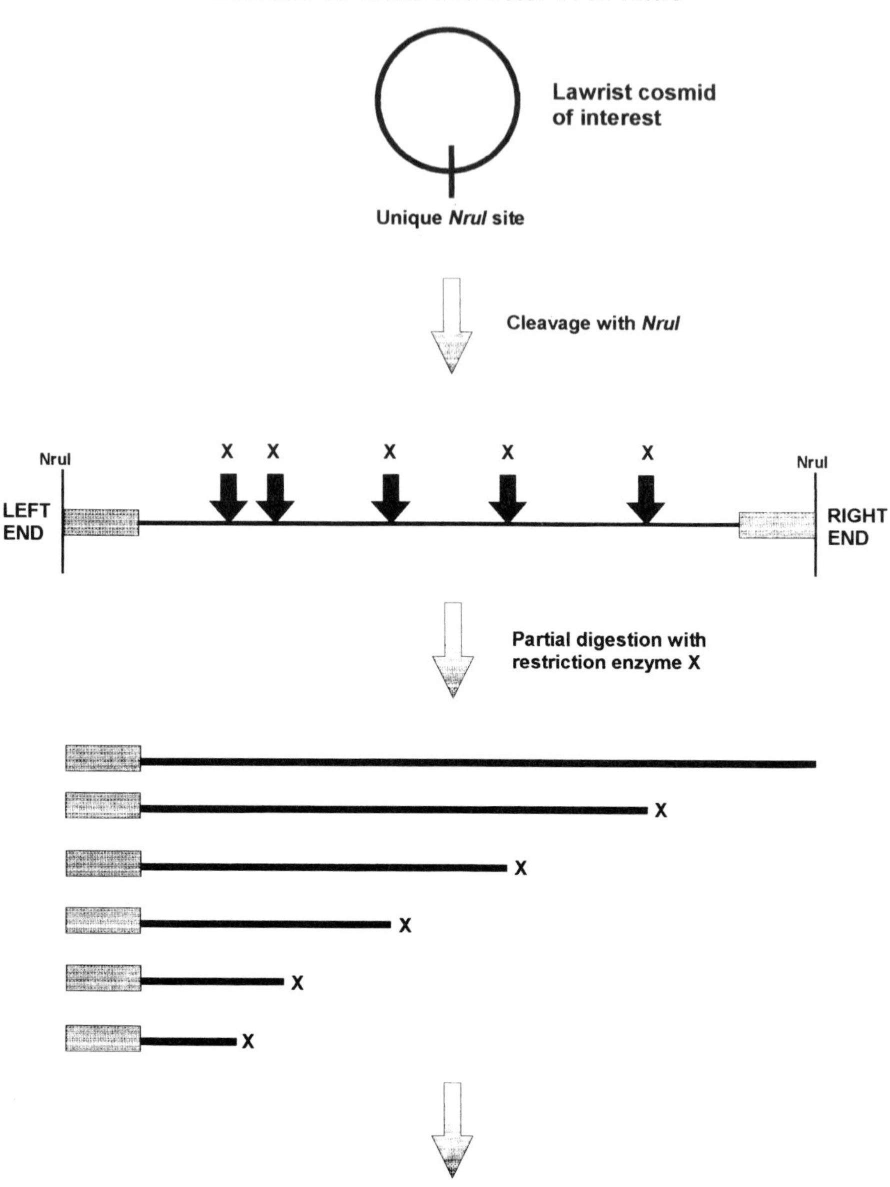

Figure 2. General method for restriction mapping a cosmid. The method shown is for cosmids made in the Lawrist series of vectors. Minor variations may be required for other vectors (see text). The cosmid DNA of interest is cleaved to completion at the unique *Nru*I site within the vector to generate a linear molecule. Partial digestion of the linear molecule with the restriction enzyme of interest (X, sites indicated by the bold arrows) generates a family of fragments. After separation by electrophoresis in an agarose gel, the frag-

restriction mapping analysis of cosmid clones is that of size; there are too many sites for most commonly used enzymes and this precludes rapid mapping. *Cos* mapping overcomes this to a degree by using partial digestion to identify map order and sites. The process used to involve cleavage of the *cos* site present in the cosmid with bacteriophage lambda terminase enzyme (38). The left or right cohesive end was labelled by annealing a ^{32}P-labelled 12 base DNA fragment that is complementary to the left or right sticky end of phage λ. Linearized DNA was then partially digested and radioactive fragments, by definition starting at the labelled end, were detected by autoradiography after size fractionation on agarose gels. Unfortunately, terminase is no longer commercially available. However a method utilizing the same principle involves linearization of the cosmid insert, followed by hybridization and partial digestion. A generalized procedure is shown schematically in *Figure 2*.

Nowadays, the most common approach to cosmid mapping is the hybridization of end-labelled primers specific to either side of the insert cloning site. After complete cleavage with a specific restriction enzyme (e.g. *Not*I for pWE15), a partial digestion series is performed with the restriction enzyme of interest (similar to that used in *Protocol 4*). Digestion products are subsequently separated by standard gel electrophoresis in 0.5% agarose gels or by FIGE (suitably modified *Protocol 17* and *Protocol 18*). After transfer to a nylon membrane, sequential hybridization with T3 and T7 sequencing primers radiolabelled using polynucleotide kinase will yield an autoradiograph of hybridizing fragments that can be used to determine the sequence of fragments from either end, yielding a concordant map. The only possible caveat against using this method is that there may be a small restriction fragment close to the hybridization site which might be lost to analysis.

An alternative method similar to the *cos* mapping procedure is to identify a unique site in the cosmid vector (e.g. *Nru*I for the Lawrist series of vectors), and then digest with a second enzyme that will cleave some distance from the unique site (e.g. *Pst*I). The fragments that have both *Nru*I and *Pst*I termini can then be used as 'end' probes. Cosmid clones of interest are cleaved to completion with *Nru*I, and a partial digestion series set-up with the restriction enzyme of interest (similar to that used in *Protocol 4*). After electrophoresis and transfer to a nylon membrane, the products can be sequentially hybridized with the 'end' probes.

Naturally, a good set of size markers is required to cover the entire molecular weight range. We usually make a concoction of size standards. This comprises

---

ments are transferred to a nylon membrane and hybridized with an 'end' probe derived from vector sequences (hatched box). Only those fragments that have homology to the end probe sequence will be identified: internal X–X fragments will not be seen. By measuring the sizes of hybridizing fragments, a concordant map can be deduced. The procedure can be repeated for the other 'end' probe, or with alternative restriction enzymes of interest.

a number of digests of bacteriophage λ (*Kpn*I, *Hin*dIII, and *Eco*RI) mixed with the commercially available 1 kb ladder (New England Biolabs).

4. Cosmid fingerprinting

4.1 Overview

A technique for the rapid analysis of cosmid clones containing DNA from the region of interest involves fingerprinting. This has been used to great effect in a number of genome mapping studies, both directly (9, 39) and indirectly via the use of YACs (40). This procedure entails the generation of a series of clone 'autographs', each comprising 15–25 labelled fragments, for the construction of an overlapping clone map or contig. While the protocols for the preparation of cosmid DNA detailed above can be employed, the number of clones to be analysed often makes this approach impractical. Accordingly, we use 96-well microtitre plates and a Biomek robot workstation. Higher density plates could be used if required, though in our case, we have been limited by our model of robot workstation. A robotic workstation is not necessary for DNA isolation and fingerprinting analyses; the manipulations can easily be performed using a multichannel pipette, though obviously at a reduced throughput.

Clones of interest are inoculated into 500 μl aliquots of L broth containing the pertinent antibiotic, using deep well plates available from Beckman. After incubation with vigorous aeration at 37°C overnight, cosmid DNA is isolated from the cultures by 'microprepping' (*Protocol 20*), and then subjected to standard fingerprinting reactions (7, 8, 41, 42). Briefly, cosmid DNAs are restricted with *Hin*dIII, their termini labelled with [^{32}P]dATP using reverse transcriptase, followed by restriction digestion with *Sau*3AI (*Protocol 21*). The products are denatured by heat and subjected to electrophoresis on 4% denaturing polyacrylamide gels (*Protocol 25*) with suitable molecular length standards (lambda (λ) DNA cleaved with the restriction enzyme *Sau*3AI, *Protocol 23*). The relative mobilities of the labelled fragments are detected by autoradiography. Variations on the above methodology are possible, and are described for the analysis of the G + C-rich genome of *Leishmania major*, the research interest of one of us (A.I.). In this instance, a simplified fingerprinting procedure (*Protocol 22*), followed by electrophoresis on a non-denaturing polyacrylamide system (*Protocol 26*) with λ DNA cleaved with *Hin*fI as molecular length standards (*Protocol 24*).

Data analysis entails scanning the autoradiograph, and importing the image thus generated into the Image software package. This runs on a Sparc/Silicon Graphics Workstation, and performs the lane- and band-finding algorithms, using software, originally written by John Sulston (jes@sanger.uk), that has been re-written in C by Friedemann Wobus (fw@sanger.uk) and Richard Durbin (rd@sanger.uk). This software is also available by anonymous ftp from rona.sanger.uk, in the /pub/image directory. At the time of writing, the version number is 2.1, and comes complete with MAN (manual) pages. An in-depth description of the functionality of the software packages available is beyond the scope of this article (the authors of the software should be consulted for more information). The data that Image generates can be directly imported into ACEDB-like relational databases (contact Richard Durbin (rd@sanger.uk) or Jean Thierry-Mieg (mieg@kaa.crrn-mop.fr) for more details).

4.2 Procedures

Cosmid DNAs are isolated in large numbers using microtitre plates. The procedure used is essentially a scaled down version of a standard 'mini-prep', and yields consistently high quality DNA that can be used for a number of analyses, including fingerprinting. The electrophoretic separation of the fingerprinting reactions is a critical step: the fingerprints are analysed by computer and this requires rigid adherence to certain procedures. Accordingly, polyacrylamide gels are bonded to one of the glass plates, resulting in the high dimensional stability that is required for absolute migration distances to be accurate.

A specific lane 'architecture' is required: the comb used for polyacrylamide gel electrophoresis must have lanes 3 mm wide with an interlane separation of 2 mm. A comb manufactured by IBI meets these specifications. There must be uniform spacing across the loaded tracks (two smaller combs can not be used instead of a single full-width one). Shark's tooth combs can not be used (the software can not identify lanes if there is no gap between them). The order of samples to be run on the gel is critical to the final data analysis. The first track **must** be a marker followed by six samples and then a marker followed by six samples and so on across the gel ending with a marker track.

Bacteriophage lambda DNA restricted with *Sau*3AI (for denaturing polyacrylamide gel electrophoresis) or *Hin*fI (for non-denaturing systems) and subsequently end-labelled with [^{35}S]dATP is used as a molecular weight marker for gel electrophoresis. These DNA standards are used by the computer software to normalize for gel-to-gel electrophoretic variation: without high quality markers (no partials, clean bands, no gel artefacts), the data on any individual gel is unusable. As the entire procedure of data digitization and analysis is dependent on the markers, their quality is critical.

The computerized analysis of fingerprints and their subsequent assembly into contigs relies on a critical criterion: that all data can be directly compared. This is achieved by process of normalization. In outline, the fingerprint autoradiographs are scanned and the position of fragments detected by alterations in the optical density of the film. The variation in densities (bands) detected by the scanner are converted into an image of the autoradiograph as a whole. The software attempts to find lanes within this image. Assuming that the markers are of sufficiently high quality, the first and last lanes are identified, both known to be markers, and every seventh track similarly identified as a marker track. Software then attempts to align ('lock') an internal image of a standardized marker set on the experimental markers, i.e. those on the autoradiograph. The expansion or contraction of the experimental markers compared to the standard set is calculated, enabling a direct measure of gel mobility compared to the ideal standardized marker set. An image of the gel with locked markers and sample bands indicated on it is sent to the terminal screen. At this stage, manual editing of the clone fingerprint image can be undertaken. The editing facility enables the user to manually monitor the entire normalization process, and allows the entry of suboptimal autoradiographic data, as the software is able to compensate for 'smiling' gels. When editing is complete, the information is stored: the name of the clone, the number of the gel, the number of fragments it contained, and their normalized mobility. In this fashion, migration distances of clones analysed on different gels can readily be compared to within 0.7 mm.

Protocol 20. Micropreparation of DNAs

Equipment and reagents

- Solution I: 50 mM glucose, 10 mM EDTA, 25 mM Tris pH 8.0
- Solution II: 0.2 M NaOH, 1% SDS
- Solution III: 3 M Na acetate pH 5.5
- 5 M LiCl
- Plate sealers (ICN Biomedicals, 7740005)

Method[a]

1. Transfer 250 µl of the overnight cultures into a microtitre plate. Seal the plate with a polythene plate sealer and centrifuge at 2500 r.p.m. (930 *g*) in a bench-top centrifuge for 2 min. (A Dupont T6000B centrifuge, with H1000(B) rotor and microtitre plate carriers, is suitable for this purpose.) Discard the supernatant, draining the microtitre plate briefly on absorbent tissues.
2. Resuspend the pellets as much as possible in the small amount of L broth that remains in the wells by gently tapping the side of the microtitre plate. Add 25 µl of solution I. Add 50 µl freshly made solution II. Mix by tapping the tray, and leave for 5–10 min at room temperature, by which time the solution should have cleared. Finally, add 25 µl of cold solution III. Seal the plates, and invert several times until a precipitate has formed. After incubation at 4°C for 5 min, centrifuge the plate at 3500 r.p.m. for 10 min at 4°C.
3. Dispense 55 µl of isopropanol into another microtitre plate. Transfer 90 µl of the supernatant from step 2 into the plate containing isopropanol, seal, and mix thoroughly.
4. Precipitate the nucleic acids by placing the plate at 4°C for 1 h, and pellet by centrifugation at 3500 r.p.m. for 10 min at 4°C. Discard the

supernatant and drain the plate on absorbent tissues. Add 25 μl of sterile water to the pellets, followed by 25 μl of cold 5 M LiCl.

5. Seal the plate, mix thoroughly, and leave at 4°C for at least 1 h. Centrifuge the plate at 3500 r.p.m. for 10 min at 4°C to pellet the debris. While the plate is spinning, dispense 100 μl of 99% ethanol into another plate. Transfer the entire supernatant (50 μl) to the ethanol plate. Seal the plate, mix thoroughly, and incubate the plate at −20°C until required.
6. Recover purified cosmid DNA by centrifugation at 3500 r.p.m. for 10 min at 4°C, discarding the supernatant. Gently wash the pellets by the addition of 150 μl of 70% ethanol, followed by re-centrifugation. Gently discard the supernatant, drain briefly, and allow to air dry for a few minutes at room temperature. Resuspend the pellets in 10 μl (Lorist cosmids) or 40 μl (sCos1, pWE15) of TE pH 8.0.

[a] All volumes referred to are per well of a 96-well microtitre plate.

Protocol 21. Fingerprinting of cosmid DNAs using *Hin*dIII and *Sau*3AI [a]

Equipment and reagents

- AMV reverse transcriptase (Northumbria Biologicals Limited, 020604)
- 10 × M buffer: supplied with the *Hin*dIII restriction enzyme
- *Hin*dIII (Boehringer Mannheim, 656321)
- *Sau*3AI (Boehringer Mannheim, 709751)
- ddGTP (Pharmacia, 27207501)
- [^{32}P]dATP (Amersham International, PB10384)
- Plate sealers (ICN Biomedicals, 7740005)
- Formamide stop dye: deionized formamide, 0.03% (w/v) bromophenol blue, 0.03% (w/v) xylene cyanol, 10 mM EDTA pH 8.0

Method

1. Make up fingerprint reaction mixture sufficient for one full microtitre tray (96 samples) as follows:
   - 10 × M buffer 40 μl
   - 1 mm ddGTP 5 μl
   - sterile double distilled water 160 μl
   - AMV reverse transcriptase (10 U/μl) 4 μl
   - *Hin*dIII (10 U/μl) 4 μl
   - [^{32}P]dATP 8 μl
2. Transfer 2 μl of the DNA generated by *Protocol 20* to a fresh microtitre plate on ice. Try to dispense the DNA solution on to the bottom of each well.
3. Add 2 μl of the reaction mix detailed in step 1 to the side of each well in the tray, using a Hamilton Syringe or multichannel pipette.

Protocol 21. *Continued*

4. Seal, and briefly centrifuge the plate to ensure thorough mixing of the reaction components. Incubate the plate at 37°C for 45 min, followed by incubation at 68°C for 25 min. The latter incubation is to heat inactivate the reverse transcriptase and *Hin*dIII. Cool the plate on ice for 5 min and recover any condensation that has formed by a brief centrifugation.
5. Add 2 μl of the following *Sau*3AI mixture to the side of each well:
   - 10 × M buffer 22.5 μl
   - sterile double distilled water 197 μl
   - *Sau*3AI (50 U/μl) 5.7 μl
6. Centrifuge briefly, and incubate at 37°C for 2 h.
7. After incubation, add 4 μl of formamide stop dye, re-centrifuge, and store at −20°C.

[a] This method is based on refs 9 and 43.

Protocol 22. Fingerprinting of cosmid DNAs using *Hin*fI[a]

Equipment and reagents

- Klenow polymerase (Boehringer Mannheim, 1008412)
- 10 × H buffer: supplied with *Hin*fI enzyme
- *Hin*fI (Boehringer Mannheim, 779679)
- dGTP, dATP, dTTP (Pharmacia, 27203501)
- BSA (Sigma, B2518)
- RNase A (Sigma, R9009)
- [^{32}P] dCTP (ICN Biomedicals, 33004X)
- Plate sealers (ICN Biomedicals, 7740005)
- Standard gel loading dye mix (5 ×): 0.25% bromophenol blue, 0.25% xylene cyanol, 15% Ficoll 400

Method

1. Make up fingerprint reaction mixture sufficient for two full microtitre trays (192 samples) as follows:
   - 10 × H buffer 100 μl
   - sterile double distilled water 930 μl
   - 2 mM dGTP 20 μl
   - 2 mM dATP 20 μl
   - 2 mM dTTP 20 μl
   - 10 mg/ml BSA 20 μl
   - 10 mg/ml RNase A 4 μl
   - Klenow polymerase (5 U/μl) 4 μl
   - *Hin*fI (10 U/μl) 60 μl
   - [^{32}P]dCTP 4 μl

2. Transfer 4 μl of the DNA generated by *Protocol 20* to a fresh microtitre plate on ice. Try to dispense the DNA solution into the bottom of each well.
3. Add 6 μl of the reaction mix detailed in step 1 to the side of each well in the tray, using a Hamilton Syringe or multichannel pipette.
4. Seal, and briefly centrifuge the plate to ensure thorough mixing of the reaction components. Incubate the plate at 37°C for 90 min.
5. After the incubation, add 4 μl of standard gel loading dye mix, re-centrifuge, and store at −20°C.

[a] This method is based on that of Knott *et al.* (44).

Protocol 23. Lambda marker DNA: a *Sau*3AI digest suitable for denaturing gel electrophoresis of fingerprints generated by *Hin*dIII and *Sau*3AI digestion

Reagents

- *Sau*3AI (Boehringer Mannheim, 709751)
- 10 × M buffer: supplied with *Sau*3AI enzyme
- Unmethylated lambda DNA (Gibco–BRL, 5205250SA)
- [^{35}S]dATP (Amersham International, SJ264)
- AMV reverse transcriptase (Northumbria Biologicals Limited, 020604)
- dGTP (Pharmacia, 27203501)
- ddTTP (Pharmacia, 27483101)
- Formamide stop dye: deionized formamide, 0.03% (w/v) bromophenol blue, 0.03% (w/v) xylene cyanol, 10 mM EDTA pH 8.0

Method

1. Digest 10 μg of lambda DNA with *Sau*3AI for 1 h at 37°C in the following reaction:
   - lambda DNA (10 μg) 25 μl
   - 10 × M buffer 50 μl
   - sterile double distilled water 415 μl
   - *Sau*3AI (50 U/μl) 10 μl
2. Label an aliquot of the digested DNA at 37°C for 30 min in the following reaction mixture:
   - *Sau*3AI cut lambda DNA 43.5 μl
   - 10 × M buffer 1.1 μl
   - 10 × mM dGTP 2 μl
   - 10 × mM ddTTP 2.5 μl
   - [^{35}S] dATP 4 μl
   - AMV reverse transcriptase (10 U/μl) 1 μl

Protocol 23. *Continued*

3. Place the remaining DNA from step 1 at −20 °C without otherwise terminating the reaction.
4. The end-filling reaction (step 2) is terminated by the addition of an equal volume of formamide stop dye.[a] Analyse 2 μl by electrophoresis on a 4% denaturing polyacrylamide gel and autoradiography (*Protocol 25*). On the basis of the autoradiograph obtained, make a suitable dilution in formamide stop dye (if required) of the labelled marker. Store at −20 °C.

[a] We have observed a decrease in quality of the marker preparation over time, presumably as a result of freeze–thaw cycles in the presence of formamide. For this reason, if the labelled marker is not going to be used immediately, we recommend that labelled DNA is stored in the absence of formamide, and that formamide stop dye is added prior to boiling and electrophoresis.

Protocol 24. Lambda marker DNA: a *Hin*fI digest suitable for non-denaturing gel electrophoresis of fingerprints generated by *Hin*fI

Reagents

- Klenow polymerase (Boehringer Mannheim, 1008412)
- *Hin*fI (Boehringer Mannheim, 779679)
- 10 × H buffer: supplied with *Hin*fI enzyme
- Unmethylated lambda DNA (Gibco–BRL, 5205250SA)
- [^{35}S]dATP (Amersham International, SJ264)
- dGTP, dTTP, dCTP (Pharmacia, 27203501)
- Standard gel loading dye mix (5 ×): 0.25% bromophenol blue, 0.25% xylene cyanol, 15% Ficoll 400

Method

1. Digest 10 μg of lambda DNA with *Hin*fI for 1 h at 37 °C in the following reaction.
   - lambda DNA (10 μg) 25 μl
   - 10 × H buffer 50 μl
   - sterile double distilled water 405 μl
   - *Hin*fI (10 U/μl) 20 μl
2. Label an aliquot of the digested DNA at room temperature for 30 min using the conditions detailed below:
   - *Hin*fI cut lambda DNA 43.5 μl
   - 2 mM dCTP 5 μl
   - 2 mM dGTP 5 μl
   - 2 mM dTTP 5 μl
   - [^{35}S]dATP 4 μl
   - Klenow polymerase (5 U/μl) 1 μl

3. Store the remainder of the digested DNA from step 1 unterminated at −20°C.
4. Dilute a small aliquot of the labelling reaction (step 2) fifteenfold with TE pH 8.0. Add one-fifth volume standard gel loading mix, and analyse 2 μl by electrophoresis on a 4% non-denaturing polyacrylamide gel and autoradiography (*Protocol 26*). On the basis of the autoradiograph obtained, make a suitable dilution of the remainder of the labelled marker. Store at −20°C. No obvious deterioration of marker quality has been observed over time (see footnote to *Protocol 23*).

Protocol 25. Denaturing polyacrylamide gel electrophoresis

Reagents

- Dimethyldichlorosilane (Repelcote) (Merck, 63211645)
- Bonding solution: 3 ml ethanol, 50 μl 10% acetic acid, 2.5 μl of WACKER Silicone solution (Wacker-Chemie GmbH, Munich, Germany)
- Urea
- Acrylamide: 40% stock (w/v) (19:1 acrylamide/bisacrylamide)
- 10% ammonium persulfate
- TEMED (*N,N,N′,N′*-tetramethylethylenediamine) (Sigma, T8133)

Method

1. Standard BRL sequencing gel tanks (Model S2) are used for the electrophoretic analysis of fingerprinting reactions.[a] Apply siliconizing agent (dimethyldichlorosilane) to the larger plate, spread evenly, and leave for 10 min. Rinse the plate with water and wipe dry.
2. Care is required when applying the bonding solution to the smaller gel plate, as 'over-bonding' will result in the gel sticking to both plates and splitting when the gel is processed. Wash the smaller plate well with detergent, dry, wipe with ethanol, and then wipe with the bonding solution. Leave the plate for 1 min after applying the bonding solution, and then wash the plate with ethanol. Tape the plates as normal.
3. For each gel, dissolve 42 g of urea in 45 ml of distilled water plus 10 ml of 10 × TBE. When the urea has dissolved, add 10 ml of the 40% acrylamide stock, 800 μl of 10% ammonium persulfate, and 80 μl of TEMED. Mix well for 10 sec and pour the gel. After pouring, insert the comb and clamp the plates around the comb. Leave the gel to set for 1 h in an almost horizontal position. Gels can be prepared the day before they are used and wrapped in Saran wrap to prevent drying.
4. Prior to electrophoresis, the combs are removed and sealing tape stripped off the bottom of the plates to enable the migration of the bromophenol dye to be observed. 1 × TBE is used for electrophoresis.

Protocol 25. *Continued*

The wells should be carefully flushed to remove bubbles and debris prior to sample loading. Pre-running the gel does not appear to be necessary. Denature the fingerprint reactions in an oven at 80°C for 10 min. Markers (*Protocol 23*) are denatured by boiling for 5 min. The urea that has leached out of each set of six wells (half a row of a microtitre plate) should be flushed out before loading the samples (usually 2 μl). The markers (2 μl) are then loaded, every seventh lane, eight to a gel. Normally, 42 cosmid fingerprints are run per gel.

5. Gels are run at constant power (74 W) until the bromophenol blue dye front is 1 cm from the bottom of the gel. Gel plates are carefully separated by removing the tape and prising the plates apart at one corner with a scalpel. The gel sticks to the bonded plate, which is first placed in a large tray containing 1 litre of 10% acetic acid for 15 min, and subsequently in water for 20 min prior to drying in an oven (100°C) for 45 min.
6. Wrap the dried gel (while it is still hot) in Saran wrap, squeezing out the air bubbles with a tissue. Clean another glass plate. Place this over the gel plate and hinge it on one side with tape. Place a 35 cm × 43 cm sheet of Kodak XAR5 film or similar between the plates with the film overhanging the wells. Expose for one to three days at room temperature in an envelope, taking care not to disturb the plates. When developing the autoradiograph, be careful to mark the film on one corner to ensure that the orientation of the sample loading is known.

[a] The use of sequencing gel apparatus other than the BRL S2 is not encouraged. The computerized analysis of fingerprint autoradiographs generated in step 6 requires certain criteria to be met; one of these is the physical dimensions of the autoradiographs (35 cm × 43 cm), and the data on them. Autoradiographs are scanned using an Amersham gel scanner; the data for each sample, whether cosmid clone or marker track, must include the well into which the sample was loaded.

Protocol 26. Non-denaturing polyacrylamide gel electrophoresis[a]

Reagents

- Dimethyldichlorosilane (Repelcote) (Merck, 6321645)
- Bonding solution: 3 ml ethanol, 50 μl 10% acetic acid, 2.5 μl of WACKER Silicone solution (Wacker–Chemie GmbH, Munich, Germany)
- Acrylamide: 40% stock (w/v) (19:1 acrylamide/bisacrylamide)
- 10% ammonium persulfate
- TEMED (*N,N,N′,N′*-tetramethylethylenediamine) (Sigma, T8133)

Method

1. Proceed as for *Protocol 25*, steps 1 and 2.

2. For each gel, mix 10 ml of 10 × TBE and 10 ml of the 40% acrylamide stock with 80 ml distilled water. Add 800 μl of 10% ammonium persulfate and 80 μl of TEMED. Mix well for 10 sec and pour the gel. After pouring, insert the comb and clamp the plates around the comb. Leave the gel to set for 1 h in an almost horizontal position. Gels can be prepared the day before they are used and wrapped in Saran wrap to prevent drying.
3. Prior to electrophoresis, the combs are removed and sealing tape stripped off the bottom of the plates to enable the migration of the bromophenol dye to be observed. 1 × TBE is used for electrophoresis, and the wells should be carefully flushed prior to sample loading to remove bubbles and debris. The samples (usually 20 μl) are loaded in groups of six (half a row of a microtitre plate). The markers (20 μl) are then loaded, every seventh lane, eight to a gel. Normally, 42 cosmid fingerprints are run per gel.
4. Gels are run at constant power (40 W) until the bromophenol blue dye front is 7 cm from the bottom of the gel. Gel plates are carefully separated by removing the tape and prising the plates apart at one corner with a scalpel. The gel sticks to the bonded plate, which is immediately placed, without fixing, in an oven at 120°C for 45 min.
5. Wrap the dried gel (while it is still hot) in Saran wrap, squeezing out the air bubbles with a tissue. Clean another glass plate. Place this over the gel plate and hinge it on one side with tape. Place a 35 cm × 43 cm sheet of Kodak XAR5 film or similar between the plates with the film overhanging the wells. Expose for one to three days at room temperature in an envelope, taking care not to disturb the plates. When developing the autoradiograph, be careful to mark the film on one corner to ensure that the orientation of the sample loading is known.

[a] This method is based on that of Knott *et al.* (44).

5. Conclusion

In this article, we have detailed the construction of cosmid libraries. The cosmids generated by these methods can either be used as resources for further analyses without any attempts to establish their relationships directly (reference clones), or they can be aligned into contigs to make a cloned DNA map. This latter area has only been briefly discussed and further information can be obtained by contacting the authors directly by e-mail. A large project that relies on fingerprint analysis is not a trivial proposition to set-up successfully. Fortunately, this is now a popular approach to the study of genomes as the technology continues to develop.

References.

1. Buxton, J. P., Shelbourne, P., Davies, J., Jones, C., Perryman, M. B., Ashizawa, T., *et al.* (1992). *Genomics*, **13**, 526.
2. Buckler, A. J., Chang, D. D., Graw, S. L., Brook, J. D., Haber, D. A., Sharp, P. A., *et al.* (1991). *Proc. Natl Acad. Sci. USA*, **88**, 4005.
3. Huntington's Disease Collaborative Research Group. (1993). *Cell*, **72**, 971.
4. Sulston, J., Du, Z., Thomas, K., Wilson, R., Hillier, L., Staden, R., *et al.* (1992). *Nature*, **356**, 37.
5. Dujon, B., Alexandraki, D., Andre, B., Ansorge, W., Balladron, V., Ballesta, J. P. G. *et al.* (1994). *Nature*, **369**, 371.
6. Wilson, R., Ainscough, R., Anderson, K., Baynes, C., Berks, M., Bonfield, J., *et al.* (1994). *Nature*, **368**, 32.
7. Redeker, B., Hoovers, J. M. N., Alders, M., van Moorsel, K. J. A., Ivens, A. C., Gregory, S., *et al.* (1994). *Genomics*, **21**, 538.
8. Heding, I. J. J. P., Ivens, A. C., Wilson, J., Strivens, M., Gregory, S., Hoovers, J., *et al.* (1992). *Genomics*, **13**, 89.
9. Coulson, A., Sulston, J., Brenner, S., and Kahn, J. (1986). *Proc. Natl Acad. Sci. USA*, **83**, 7821.
10. Kohara, Y., Akiyama, A., and Isono, K. (1987). *Cell*, **50**, 495.
11. Pierce, J. C., Sternberg, N., and Sauer, B. (1992). *Mamm. Genome*, **3**, 550.
12. Kim, U.-J., Shizuya, H., de Jong, P. J., Birren, B., and Simon, M. I. (1992). *Nuclei Acids Res.*, **20**, 1083.
13. Collins, J. and Hohn, B. (1978). *Proc. Natl Acad. Sci. USA*, **75**, 4242.
14. Hohn, B. and Collins, J. (1980). *Gene*, **11**, 291.
15. Becker, A. and Gold, M. (1975). *Proc. Natl Acad. Sci. USA*, **72**, 581.
16. Becker, A., Murialdo, H., and Gold, M. (1977). *Virology*, **78**, 277.
17. Sambrook, J., Fritsch, E. F., and Maniatis, T. (ed.) (1989). *Molecular cloning: a laboratory manual*. Laboratory Press, Cold Spring Harbor, New York.
18. Pouwels, P. H., Enger-Valk, B. E., and Brammar, W. J. (1986). *Cloning vectors. A laboratory manual*. Elsevier, Amsterdam.
19. Little, P. F. R. and Cross, S. H. (1985). *Proc. Natl Acad. Sci. USA*, **82**, 3159.
20. Cross, S. H. and Little, P. F. R. (1986). *Gene*, **49**, 9.
21. Gibson, T. J., Coulson, A. R., Sulston, J. E., and Little, P. F. R. (1987). *Gene*, **53**, 275.
22. Gibson, T. J., Rosenthal, A., and Waterston, R. H. (1987). *Gene*, **53**, 283.
23. Bates, P. F. and Swift, R. A. (1983). *Gene*, **26**, 137.
24. Seed, B., Parker, R. C., and Davidson, N. (1982). *Gene*, **10**, 249.
25. Speek, M., Raff, J. W., Harrison-Lavoie, K., Little, P. F. R., and Glover, D. M. (1988). *Gene*, **64**, 173.
26. Nizetic, D., Zehetner, G., Monaco, A. P., Gellen, L., Young, B. D., and Lehrach, H. (1991). *Proc. Natl Acad. Sci. USA*, **88**, 3233.
27. Evans, G. A., Lewis, K., and Rothenberg, B. E. (1989). *Gene*, **79**, 9.
28. Nelson, D. A., Ballabio, A., Victoria, M. F., Pieretti, M., Bies, R. D., Gibbs, R. A., *et al.* (1991). *Proc. Natl Acad. Sci. USA*, **88**, 6157.
29. Clarke, L. and Carbon, J. (1976). *Cell*, **9**, 91.
30. Leach, D. R. F. and Stahl, F. W. (1983). *Nature*, **305**, 448.

31. Neil, D. L., Villasante, A., and Fisher, R. B. (1990). *Nucleic Acids Res.*, **18**, 1421.
32. Strand, M., Prolla, T. A., Liskay, R. M., and Petes, T. D. (1993). *Nature*, **365**, 274.
33. Hanahan, D. and Meselson, M. (1980). *Gene*, **10**, 63.
34. Feinberg, A. P. and Vogelstein, B. (1983). *Anal. Biochem.*, **132**, 6.
35. Bentley, D. R., Todd, C., Collins, J., Holland, J., Dunham, I., Hassock, S., *et al.* (1992). *Genomics*, **12**, 534.
36. Heding, I. J. J. P. and Little, P. F. R. (1990). *Technique*, **2**, 218.
37. Rackwitz, H.-R., Zehetner, G., Murialdo, H., Delius, H., Chai, J. H., Poustka, A., *et al.* (1985). *Gene*, **40**, 259.
38. Becker, A. and Gold, M. (1978). *Proc. Natl Acad. Sci. USA*, **75**, 4199.
39. Stallings, R. L., Doggett, N. A., Callen, D., Apostolou, S., Chen, L. Z., Nancarrow, J. K., *et al.* (1992). *Genomics*, **13**, 1031.
40. Bellanne-Chantelot, C., Lacroix, B., Ougen, P., Billault, A., Beaufils, S., Bertrand, S., *et al.* (1992). *Cell*, **70**, 1059.
41. Coulson, A. and Sulston, J. (1988). In *Genome mapping: a practical approach* (ed. K. E. Davies), pp. 19–39. IRL Press, Oxford.
42. Harrison-Lavoie, K. J., John, R. M., Porteous, D. J., and Little, P. F. R. (1989). *Genomics*, **5**, 501.
43. Heding, I. J. J. P., Strivens, M., Gregory, S., Ivens, A. C., and Little, P. F. R. (1991). *Technique*, **3**, 129.
44. Knott, V., Rees, D. J., Cheng, Z., and Brownlee, G. G. (1988). *Nucleic Acids Res.*, **16**, 2601.

2

Chromosome-specific gridded cosmid libraries: construction, handling, and use in parallel and integrated mapping

DEAN NIŽETIĆ and HANS LEHRACH

1. Introduction

The analysis of the information in human genomic DNA is based on the cloning and molecular characterization of the genes together with the genetic analysis of the phenotypes associated with the genetic loci under study. Libraries of cloned genomic DNA serve as essential intermediates to relate the information from the genetic analysis of phenotypic variation to the molecular analysis of the genes (1). Such libraries can be either constructed in a yeast or bacterial vector–host system. The yeast system, using yeast artificial chromosomes (YACs) (2) as vectors, is well suited for rapid coverage of long genomic regions and entire chromosomes (3) (see also Chapter 4). This offers easier integration of the data with the genetic and physical maps of the genomic DNA, for example in the comparison of the maps of cloned DNA to long-range restriction maps of genomic DNA constructed using restriction endonucleases that cut at infrequent intervals. The yeast system is however not the best choice for the application of methods to identify transcribed regions (genes), to analyse the transcription unit and exon–intron organization, and to obtain the sequence of the genes. The longer the YAC clones are, the more frequently they suffer from deletions, rearrangements, co-cloning, and chimerism, which makes the establishment of a high-resolution map difficult. Moreover, irrespective of the YAC length, the isolation of pure YAC DNA from the genomic DNA of the host is difficult to achieve in most cases (see Chapter 4 for a discussion of these problems). The bacterial (*E. coli*)-based vector systems (phage, cosmid, P1, F-) (4–6) are much more suitable for the characterization of transcribed sequences (7–10), to study the molecular organization of the genes (11), and to provide material for DNA sequencing. The isolation of the cloned DNA in a pure state, free from the DNA of the host, is much more straightforward. Cosmid libraries are probably the simplest to use. They can be constructed with very high efficiency, allowing libraries to be made from limited amounts of DNA, and have been in use for longer than the P1, PAC, and other new bacterial vector systems.

A feature that makes cosmids well suited for high-resolution mapping is the relatively short length of DNA that can be cloned within them (intermediate between plasmids and YACs). This does however cause a practical problem: the very large number of clones that need to be screened. Thus a fivefold coverage of a typical mammalian genome in cosmids is represented by close to 400 000 clones. YAC clones, with 10–30-fold longer inserts, in contrast, require correspondingly fewer clones to be screened. Among other factors, this makes the use of chromosome-specific libraries especially attractive if the chromosomal assignment of the probe is known. In the case of the human genome, the use of such libraries can reduce the number of clones to be screened by 10–30-fold.

In addition, the availability of cosmid libraries enriched for specific chromosomes enables specific subsets of genes to be identified. This is especially useful in the case of large gene families distributed over multiple chromosomes, and can aid in the selective identification of sequences corresponding to segments of chimeric YACs. To allow the use of chromosome-specific cosmid libraries, we have therefore constructed a number of libraries from flow sorted chromosomes. Techniques used in the library construction will be covered in the first section of this chapter.

The following sections describe a set of techniques which enable:

(a) Permanent storage of all individual clones of the library.

(b) The parallel use of the same library by many research teams at the same time.

(c) The use of the cosmid library both as a high-resolution mapping tool and a clone resource.

(d) The incorporation of chromosome-specific cosmid clones in a relational mapping system integrating mapping data obtained through hybridization from different cloning systems and different aspects of mapping.

Since genomic cosmid libraries are created by partial digestion of genomic DNA at very frequently occurring sites, the end-points of cosmid inserts can be considered to be randomly distributed. That means that a genomic locus cloned from two different libraries will be represented by cosmids bearing different but overlapping insert sequences. A much more efficient use of the libraries is the 'reference library' system (1), by which one central cosmid library for a single chromosome is gridded into microtitre plates, and specific clones become reference start and end-points of the genomic cloned fragments. This eases the cross-referencing of data. If such a library can be displayed (spotted) on to identical duplicate nylon membranes for screening, the same library can be screened by many users of the system at the same time, and data could be stored in the central reference library database. This system started four years ago at ICRF with the construction of flow sorted libraries for the human chromosomes X, 17, 21, and 22 (12). Until now these libraries have been screened by more than 170 users using over 1000 single copy and complex probes which have detected upwards of 25 000 cosmid clones (13). More recently, cosmid libraries for the chromosomes 1, 6, 7, 11, 13, and 18 have been successfully incorporated into the system (14). The libraries are displayed on hybridization membranes at high density (currently more than 20 000 clones per 22 × 22 cm membrane). The amount of cosmid DNA in each lysed colony suffices, with the protocols outlined below, for the detection of signals even with very complex probes, such as whole YACs or probes derived from large chromosomal fragments carried in hybrid cell lines.

2. Construction of cosmid libraries from flow sorted chromosomes

Chromosome-specific libraries can be constructed using two different approaches:

(a) Selection of cosmids containing human repetitive sequences from the whole genomic libraries from hybrid cell lines which carry a single human chromosome in a rodent genome background (15).

(b) Construction of cosmid libraries directly from the DNA of a single human chromosome separated from the rest of the genome by sorting in a fluorescence activated flow sorter (12).

The first approach, although technically easier, is very labour-intensive. About 400 000 cosmids have to be hybridized to total human DNA; 5–20 000 positive clones have to be scored and aligned with the membrane which contains live colonies, which can then be gridded. This approach also suffers from the unequal distribution of repetitive sequences throughout the genome. It leaves the possibility that some cosmids will not be identified. Furthermore, some repetitive sequences in the rodent genome cross-hybridize with the human DNA, resulting in the contamination of the library with cosmids bearing rodent DNA. The second approach gives much purer libraries, and theoretically has the capability of cloning all clonable sequences from the chromosome (12, 16), but does suffer from cross-contamination of the chromosomes enriched by flow sorting.

In this section we will give a set of protocols describing how to proceed from a suspension of flow sorted chromosomes, to a cosmid library. The use of this approach requires care in the handling of very small quantities of chromosomal DNA, unless very large numbers of chromosomes can be sorted (at correspondingly increased cost and effort). The use of this approach depends on the availability of a cell line that gives a flow-karyotype profile in which a particular chromosome can be resolved from the other chromosomes. The flow sorting step as such has been covered in detail in a number of manuals and textbooks (e.g. ref. 17), and will not be discussed in detail here. In brief, several millions of cells of a proliferating cultured cell line are arrested in metaphase by the addition of colcemid. Gentle protocols are applied to lyse the nuclear membrane which releases the metaphase chromosomes into a suspension buffer in which they are stained with one or more fluorescent dyes. The target chromosome is then identified by its pattern of fluorescence emission, using either a single dye in a one laser system, or two dyes in a two laser detection system. Millions of copies of the specific chromosome are then accumulated in a test-tube.

2.1 Preparation of DNA

DNA is prepared from sorted chromosomes as described in *Protocol 1*.

Protocol 1. Preparation of DNA from flow sorted chromosomes

Reagents

- TEN: 10 mM Tris–HCl pH 8.0, 1 mM EDTA, 100 mM NaCl
- Yeast tRNA (10 mg/ml stock) (Sigma)
- Sodium dodecyl sarcosine (100 mg/ml stock) (NaDodSarc, Sigma)
- EDTA (0.5 M stock)
- Proteinase K (50 mg/ml stock)
- Phenylmethylsulfonyl fluoride (PMSF, Sigma) (4 mg/ml fresh solution in ethanol)
- Ethanol
- Sodium acetate pH 6 (3 M stock)
- TE: 10 mM Tris–HCl pH 8.0, 1 mM EDTA

Method

1. To suspensions of freshly sorted chromosomes in TEN buffer containing 25 μg/ml yeast tRNA add NaDodSarc to a final concentration of 10 mg/ml, EDTA to a final concentration of 25 mM, and proteinase K to a final concentration of 0.2 mg/ml. Rock gently overnight at 50°C.
2. Inactivate the proteinase K by the addition of PMSF to the final concentration of 40 μg/ml. Incubate at room temperature for 40 min.
3. Concentrate the DNA in aliquots of 100–125 ng of chromosomal DNA (the equivalent of about 5–6 $\times$ 10^5 human X chromosomes) in 1 ml of the solution from step 2. Split to four equal aliquots and precipitate with by adding sodium acetate (pH 6.0) to 300 mM and 2.5 vol. of ethanol. Incubate overnight at −20°C. Centrifuge the pellets in a microcentrifuge for 15 min at full speed (14000 r.p.m.). Allow to air dry for 2–3 min and dissolve the DNA in 4 μl TE buffer. Rotate on a slow rotating wheel for 3 h at 4°C. This DNA is ready for the partial digestion as in *Protocol 4*.

2.2 Preparation of the cosmid vector

The Lawrist series of cosmid vectors (18) (Pieter de Jong, personal communication) derived from Lorist (19), contain some very useful features including:

(a) A phage lambda origin of replication (to yield a moderately high copy number, particularly in colonies).
(b) Terminators of transcription flanking the cloning site (to reduce the possibility of transcripts from within the human inserts interfering with the replication of the cosmid).
(c) A double '*cos*' sequence (to enhance the efficiency of cloning) (20).
(d) SP6 and T7 RNA polymerase promoters directly flanking the cloning site (to facilitate the generation of end probes for walking).
(e) Rare cutter restriction endonuclease sites flanking the cloning site (to simplify the generation of restriction map and sequencing).
(f) The absence of homology (in its recombinant form) to the common YAC vector pYAC4 (to permit direct hybridization of whole YACs to cosmids and vice versa).

Protocol 2. Preparation of the cosmid vector arms[a]

Reagents

- Lawrist4 DNA
- 10 × high salt restriction endonuclease buffer (HSRE): 500 mM Tris–HCl pH 7.5, 100 mM $MgCl_2$, 1500 mM NaCl
- 1 × low salt restriction endonuclease buffer (LSRE): 50 mM Tris–HCl pH 7.5, 10 mM $MgCl_2$
- Distilled water
- *Sca*I restriction endonuclease (10 U/μl stock)
- *Bam*HI restriction endonuclease (10 U/μl stock)
- Phenol (distilled and saturated with Tris–HCl pH 8.0)
- Chloroform/isoamyl alcohol (v/v) 24:1
- TE: 10 mM Tris–HCl pH 8.0, 1 mM EDTA
- Ether (water-saturated diethyl ether)
- 3 M sodium acetate pH 6
- Absolute ethanol
- 70% ethanol
- Calf intestine alkaline phosphatase (CIAP, Boehringer, 1 U/μl)
- 150 mM nitrilo-triacetic acid (NTA)

Method

1. Cleave the cosmid vector on a unique site between the two *cos* sites. In the case of Lawrist4, mix 20 μg of DNA, 10 μl 10 × HSRE, and make up to 91 μl with distilled water. Add 9 μl *Sca*I and incubate 2 h at 37 °C. Remove a 5 μl aliquot, add 45 μl TE, and extract once with phenol, once with chloroform/isoamyl alcohol. Extract the aqueous phase twice with ether. Add 1/10 vol. of 3 M sodium acetate and 2.5 vol. of ethanol. Leave overnight at −20 °C and concentrate by centrifugation in a microcentrifuge for 15 min at full speed. Wash the pellet once with 70% ethanol and allow to air dry. Redissolve the DNA in 5 μl TE.

Protocol 2. *Continued*

2. To dephosphorylate the *Sca*I termini, dilute the bulk reaction to 370 μl with 1 × LSRE. Add 35 μl (35 U) of CIAP and incubate for 45 min at 37 °C.
3. Add 45 μl of NTA (to a final concentration of 15 mM). Incubate for 25 min at 68 °C. Extract once with an equal volume of phenol and once with an equal volume of chloroform/isoamyl alcohol, each time re-extracting the organic phase with an equal volume of TE. Extract the aqueous phase twice with ether. Precipitate the DNA as described in step 1, and finally resuspend the pellet in 160 μl TE. Remove a 5 μl aliquot.
4. To generate cosmid arms digest with a restriction endonuclease that cleaves at the cloning site. Add 20 μl × HSRE and 15 μl *Bam*HI (150 U). Incubate for 90 min at 37 °C. Extract with phenol, chloroform/isoamyl alcohol, and ether as described in step 3. Precipitate the DNA as described in step 1. Finally redissolve the DNA in 100 μl 1 × TE. Anticipate a loss of about a quarter of the initial DNA. The vector arms should now have a concentration of about 150 ng/μl. Remove a 5 μl aliquot. The vector arms are now ready and should be stored in aliquots at −70 °C.

[a] This protocol is similar to *Protocol 8* in Chapter 1.

Protocol 3. Test of the quality of the preparation of the vector arms

Reagents

- 10 × ligase buffer: 400 mM Tris–HCl pH 7.6, 100 mM $MgCl_2$, 10 mM dithiothreitol
- 8 mM ATP
- T4 DNA ligase (New England Biolabs 100 U/μl)
- 0.7% agarose gel
- Lambda *Hin*dIII fragments as a molecular marker
- 'Gigapak Gold' *in vitro* packaging kit (Stratagene)
- *E. coli* strain ED8 767 *recA⁻rK⁺mK⁺mcrA mcrB* (21)

Method

1. To each 5 μl aliquot from steps 1, 3, and 4 of *Protocol 2* add 11 μl TE, 2 μl 10 × ligase buffer, and 1 μl ATP. Divide each reaction mixture into equal portions. Add to one member of each pair 0.5 μl of T4 DNA ligase (50 U). Incubate both the ligated and unligated samples overnight at 15° C.
2. Analyse the reaction products on a 0.7% agarose gel. Include some Lawrist4 untreated DNA and lambda *Hin*dIII fragments as a molecular marker (*Figure 1*). A single 8.3 kb band in the *Sca*I digested DNA should concatemerize only in the undephosphorylated and self-ligated sample,

and the 1.4 kb and 6.9 kb fragments should be the only fragments visible in the final preparation of vector arms. Both should shift to higher molecular weight in the self-ligated sample. Sometimes, an additional 2.7 kb fragment is seen in Lawrist4 DNA preparations that is linearized with *Sca*I but does not contain a *Bam4*HI site and therefore will not contribute to recombinant products. This is a product of a recombination event with the host during the growth of the culture for the preparation of Lawrist4 DNA (P. Little, personal communication).

3. Self-ligate 100 ng of vector arms in a 10 μl ligation reaction using conditions as described in step 1. Package the ligation products *in vitro* using 'Gigapak Gold' following the protocol supplied by Stratagene. There should be no kanamycin resistant colonies when plating on *E. coli* ED8 767 (21) as described in *Protocol 6*. Alternatively home-made packaging extracts can be used, following protocols described previously (5).

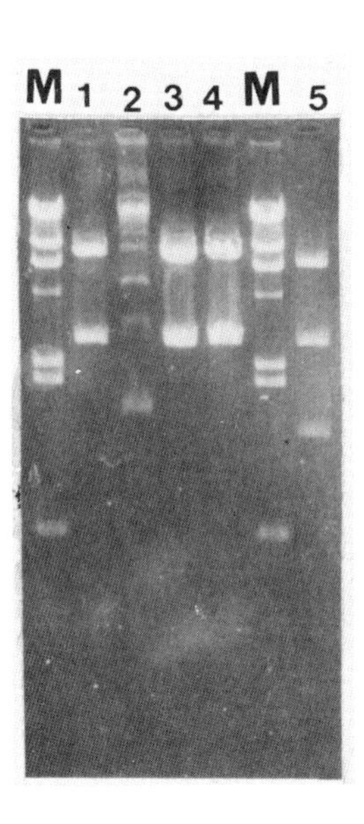

Figure 1. Gel showing the control of correctly prepared Lawrist4 vector arms (*Protocol 3*). A 0.7% agarose gel was stained with ethidium bromide and photographed under UV transillumination. Lane 1: Lawrist4 digested with *Sca*I and incubated in the ligation reaction without ligase. Lane 2: as in lane 1 but incubated with ligase. Lane 3: Lawrist4 digested with *Sca*I, dephosphorylated, and incubated in the ligation reaction without ligase. Lane 4: as in lane 3, but with ligase. Lane 5: Lawrist4 final vector arms after digestion with *Bam*HI. Lanes M were loaded with the *Hin*dIII digested phage lambda DNA as a molecular weight marker.

2.3 Partial digestion of genomic DNA

Classical protocols for the partial digestion of genomic DNA are based on the control of the extent of digestion, controlled by varying the enzyme concentration or digestion period on a fixed amount of DNA. For each particular experiment, the exact conditions have to be established empirically, as the exact concentration of genomic DNA affects the progress of the reaction. Therefore some DNA is used for the exact measurement of the concentration and the determination of the best partial digestion conditions. To minimize the losses of DNA in such reactions, an approach has been developed (22) which allows predictable levels of partial digestion using a variety of starting DNA concentrations. The approach is based on the use of a restriction endonuclease, in conjunction with a methylase that acts on the identical recognition site. Due to the competition of the two enzymes (*dam*-methylase and *Mbo*I) for the same substrate, the DNA is partially digested, the size of the resulting fragments being dependent predominantly on the ratio of the two enzymes, and to a lesser degree on DNA concentration. This makes the digestion easier to control, and has enabled us to construct libraries with extensive representation (greater than 100 000 clones) from starting amounts of approximately 200 nanograms of the flow sorted chromosomal DNA (12, 14).

Protocol 4. Partial digestions using simultaneous antagonistic enzyme protection/digestion[a,b]

Reagents

- *Dam*-methylase (New England Biolabs)
- *Mbo*I restriction endonuclease
- Enzyme diluent: 20 mM Tris–HCl pH 7.5, 100 mM KCl, 50 μg/ml gelatin, 0.1% (v/v) 2-mercaptoethanol, 50% (v/v) glycerol
- 150–300 ng concentrated chromosomal DNA
- 10 × TAK: 300 mM Tris–HCl pH 7.9, 600 mM potassium acetate, 90 mM magnesium acetate, 5 mM dithiothreitol, 800 μM S-adenosyl methionine (Biolabs), 3 mg/ml bovine serum albumin
- Calf intestine alkaline phosphatase (CIAP, Boehringer, 1 U/μl)
- 3 M NaCl
- 0.3% agarose gel
- 20–50 kb molecular weight marker: a mixture of undigested-linearized and *Hind*III digested bacteriophage lambda DNA
- TAE: 40 mM Tris–acetate pH 7.8, 1 mM EDTA
- 1μg/ml ethidium bromide in water
- 150 mM nitrilo-triacetic acid (NTA)
- Ethanol
- 70% ethanol
- TE
- Sheared total human DNA

Method

1. Prepare a fresh mixture of *dam*-methylase and *Mbo*I. Use ratios of units of *dam*-methylase: *Mbo*I of 1000, 2000, and 3000: 1. Use 0.01 U *Mbo*I for a range between 100–300 ng of chromosomal DNA. It is ideal to prepare and use fresh enzyme mixes. Alternatively they can be stored at −20°C and used within one week.
2. Prepare the partial digest of the genomic DNA by incubating 150–300 ng

DNA in a total reaction of 15 μl containing 1.5 μl 10 × TAK buffer and 1 μl of the mixture of *dam*-methylase and *Mbo*I. Carry out separate reactions with each of the mixtures from step 1. In addition, incubate a control reaction with *dam*-methylase alone and another with all other components omitting the enzymes. Incubate all samples for 3–4 h at 37 °C.

3. Dephosphorylate the termini of the genomic fragments using CIAP. Dilute the stock of CIAP 20 × with distilled water. Add 0.04–0.05 U of the freshly diluted CIAP to all digestion and control samples. Incubate for 30 min at 37 °C. Remove aliquots containing approximately 50 ng of DNA from each reaction for electrophoresis and store on ice until steps 4 and 5 are completed. Carry out electrophoresis as in step 6.
4. Inactivate the enzymes by adding NTA to a final concentration of 15 mM and incubate for 20 min at 68 °C.
5. Precipitate the DNA by adding NaCl to 250 mM, and 2.5 vol. of ethanol. Continue with step 6 and leave the DNA to precipitate overnight at −20 °C.
6. Remove aliquots for analysis by gel electrophoresis. Load approximately 50 ng of DNA from each reaction on a 0.3% agarose gel with narrow lanes (3 mm). Load one adjacent lane with the 20–50 kb molecular weight marker. Electrophorese the gel in TAE buffer (very slowly) at 0.6 V/cm for 12–16 h at room temperature. Visualize the DNA by ethidium bromide staining and UV transillumination. When the time allows, it is often helpful to Southern blot the gel, and hybridize it to total human DNA.
7. Analyse the gel picture (see the example in *Figure 2*). The first and the second bands in the lambda marker lane (48 kb and 21 kb fragments) should have separated by at least 3 cm. Untreated DNA can show a very small amount of DNA smearing below the limiting mobility. The region of limiting mobility should be well above the largest marker band. The smearing from samples not treated with *Mbo*I should not go beyond the level of the largest marker band. The digestion samples can show smearing to differing extents. Optimal digestions have smears that do not extend beyond the top of the second lambda marker band.
8. Dissolve the optimally looking digests. Centrifuge the DNA precipitates from successful partial digestions in a microcentrifuge at full speed (14 000 r.p.m.) for 15 min. Wash the pellet with 70% ethanol, allow it to air dry, and dissolve it in 5 μl TE. These samples are ready to be used in ligation (*Protocol 5*). Partial digestions prepared in this way should not be kept at 4 °C! Carry out step 8 only hours before setting-up the ligation (*Protocol 5*). Otherwise store the digests at −20 °C in ethanol.

[a] This protocol is modified from ref. 22.
[b] See also Chapter 1, *Protocol 4*; Chapter 3, *Protocol 2*; Chapter 4, *Protocol 3*.

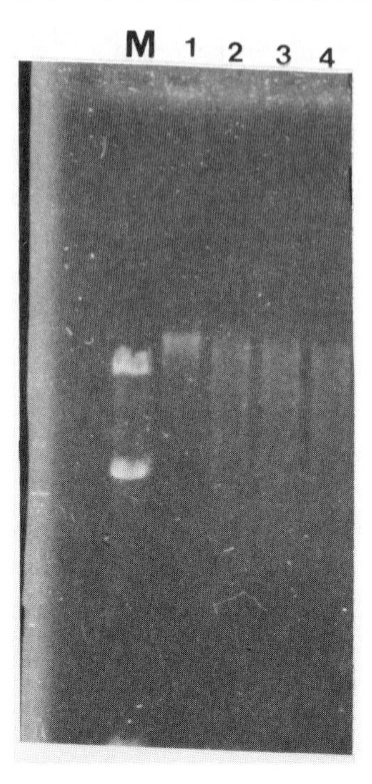

Figure 2. Gel showing an example of *dam*-methylase *Mbo*I partial digestion of the flow sorted chromosomal DNA (*Protocol 4*). A 0.3% agarose gel was stained with ethidium bromide and photographed under UV transillumination. Each lane contains 40–50 ng of chromosomal DNA which was incubated with the buffers at 37 °C without enzymes (lane 1) or with different ratios of the *dam*-methylase *Mbo*I mixture (lanes 2–4). Lane 2 is an almost optimal partial digestion for cosmid libraries. Lane M was loaded with the 20–50 kb molecular weight marker: a mixture of undigested-linearized and *Hin*dIII digested bacteriophage lambda DNA. The two prominent bands visible in this lane are the 48.5 kb (linear lambda) and the 23.1 kb (largest *Hin*dIII fragment).

2.4 Ligation, packaging, and titration

Protocol 5. Ligation and *in vitro* packaging[a]

Reagents

- 10 × ligase buffer: 400 mM Tris–HCl pH 7.6, 100 mM $MgCl_2$, 10 mM dithiothreitol
- 8 mM ATP
- T4 DNA ligase (New England Biolabs 400 U/μl)
- 'GigaPak Gold' *in vitro* packaging kit (Stratagene)
- *E. coli* strain ED8 767 *recA*$^-$ *rK*$^+$ *Mk*$^+$ *mcrA mcrB* (21)
- Lambda dilution buffer: 10 mM Tris–HCl pH 7.5, 10 mM $MgSO_4$
- 5 × SM freezing medium: 500 mM NaCl, 50 mM $MgSO_4$, 250 mM Tris–HCl pH 7.5, 10% (w/v) gelatin, 50% (v/v) glycerol

Method

1. Set-up the following reaction mixture:
   - 100 ng partially digested chromosomal DNA
   - 700 ng Lawrist4 vector arms DNA
   - 1.5 μl 10 × ligase buffer
   - 0.8 μl 8 mM ATP
   - distilled water up to 14 μl
   - 1 μl (400 U) T4 DNA ligase

   Incubate overnight at 15°C
2. Package up to 7 μl in a single packaging reaction using the protocols supplied by Stratagene for the 'GigaPak Gold II' kit.
3. In the final step, stop the packaging with the addition of 0.5 ml of lambda dilution buffer per reaction. At this stage, the reaction can be directly plated, kept for short periods at 4°C, or frozen at −70°C after the addition of 1/5 vol. of the 5 × SM freezing medium. It is best to freeze in suitable aliquots based on the titre.

[a] This protocol is similar to *Protocols 9* and *10* in Chapter 1.

Protocol 6. Plating cosmid libraries

Reagents

- *E. coli* strain DH5-α (α*mcrA* α*mcrC BsbcC*) (BRL)
- LB: 10 mg/ml bacto tryptone (Difco), 5 mg/ml yeast extract (Difco), 5 mg/ml NaCl
- 10 mM $MgSO_4$
- Agar (Difco)

Method

1. The *E. coli* strain DH5-α*mcr* is highly suitable for use as plating cells. Inoculate 40 ml LB with a fresh single colony. Culture overnight at 37°C shaking in a 100 ml conical flask. Pellet the cells by centrifugation in a medium speed centrifuge for 10 min. Resuspend gently but completely in 20 ml of sterile 10 mM $MgSO_4$. Store at 4°C. Use within four weeks.
2. Dilute the required volume of the packaged cosmids in lambda dilution buffer to a final volume of 100 μl. Mix with 100 μl of freshly resuspended cells from step 1 in microcentrifuge tubes. Incubate for 20 min at room temperature.
3. Dilute the infected cells with 1 ml LB and leave for 45 min at 37°C to enable the antibiotic resistance gene to be expressed.

Protocol 6. *Continued*

4. (a) If plating for titration on 10 cm Petri dishes, centrifuge the samples in a microcentrifuge for 2 min at 6500 r.p.m. Discard all but about 30 μl of LB. Resuspend the cells in the residual broth and spread on LB agar plates containing 30 μg/ml kanamycin.

   (b) If plating on 22 × 22 cm Nunc plates, distribute the entire 1.2 ml mixture evenly on all sides of the selective agar plates and quickly spread with a large bent arm of a Pasteur pipette to within 1 cm from the edges.

3. Gridding and handling the cosmid libraries

The following sections and accompanying protocols form a set of techniques designed to allow the permanent storage of a gridded cosmid library of easily accessible clones, and to allow the screening of the library by many different researchers (12). This approach has been applied for a range of human chromosome-specific cosmid libraries in the reference library system, where data generated from parallel studies are stored in a central relational database (13). This is a very complex operation and requires good organization, the utmost quality control upon each step, and good record keeping.

3.1 Picking colonies into microtitre plates

Colonies can be picked and 'gridded-out' either manually or automatically. The following procedure is applicable to either approach.

Protocol 7. Picking colonies into microtitre plates

Reagents

- 2 × YT: 20 mg/ml bacto tryptone (Difco), 10 mg/ml yeast extract (Difco), 5 mg/ml NaCl
- 1 × Hogness modified freezing medium (HMFM): 36 mM K_2HPO_4, 13.2 mM KH_2PO_4, 1.7 mM sodium citrate, 0.4 mM $MgSO_4$, 6.8 mM $(NH_4)_2SO_4$, 4.4% (w/v) glycerol
- 30 mg/ml kanamycin
- Ethanol
- 2 × YT broth containing kanamycin: 400 ml 2 × YT, 44 ml 10 × HMFM, 500 μl of 30 mg/ml kanamycin
- 2 × YT agar containing kanamycin (for one 22 × 22 cm Nunc dish): 200 ml 2×YT, 3 g agar (Difco), autoclave, cool to 50°C, 20 μl of 30 mg/ml kanamycin

Method

1. Plate the library that is to be picked following *Protocol 6*, but using 22 × 22 cm Nunc sterile culture dishes containing 200 ml of 2 × YT agar containing kanamycin. Cosmid libraries are ideally plated for manual picking at a density of 1000–1500 clones per 22 × 22 cm Nunc plate.[a]

2. Label microtitre plates using a black 'texta-permanent' or equivalent marker. Always label the side of the microtitre plate, never the lid.[b]
3. Add 90 μl 2 × YT broth containing kanamycin to each well of a 96-well plate, or 45 μl when using quadruple density plates (Genetix) (23). Make up one extra plate which will serve only as the control for the sterility of the procedure and which will be incubated together with the cosmid plates.
4. Pick individual colonies into individual microtitre plate wells.[c] Before culture of the picked cosmids, wrap stacks of six microtitre plates with Saran wrap on all sides and place them in an appropriate box to avoid shaking, falling, and tilting. Check the temperature in the incubator is actually 37°C. Incubate the plates and the control medium plate for 14–16 h at 37°C without shaking.
5. The following day, unwrap and inspect the medium control plate and three random plates. Vortex (carefully) three plates and observe the cloudiness of the medium. Pack the plates for freezing (see section 3.3) and freeze.

[a] The density for the automatic picking depends on the resolution and efficiency of the automatic device used. Our device routinely utilizes 2000–3000 colonies on a 22 × 22 cm Nunc plate. Plate a total number of colonies not greater than that which the team is capable of picking within two days.
[b] Make sure the type of the microtitre plates used is compatible with the gadgets on the automatic devices used in further handling. Each microtitre plate has to be labelled uniquely, showing the library, vector kind, host (if necessary), microtitre plate number, and the microtitre plate set number.
[c] Observe the plated colonies. Note any abnormalities on the plates before starting picking. This includes colonies that are too dense or too scarce, water drops from condensation on lids that could cross-contaminate colonies, different colony morphology that could indicate fungal or bacterial contamination.

It is important while picking colonies manually that the bench is well organized. The following equipment should be carefully laid out:

- wheel-pickers (an example is shown in *Figure 3A*) should be cleaned with brush and water and autoclaved at the beginning of each day
- 600 ml beaker containing 300 ml ethanol
- a Bunsen burner for flame sterilization with clear space far from combustible material adjacent to the burner (where drops of burning ethanol can fall)
- a rack for cooling the wheel-pickers

This sterilizing area has to be at the safe distance from the Nunc and microtitre plates to prevent droplets of ethanol flying into them. All other flammable material should be well removed from the area.

Stack the microtitre plates in groups of six. Keep those microtitre plates

A

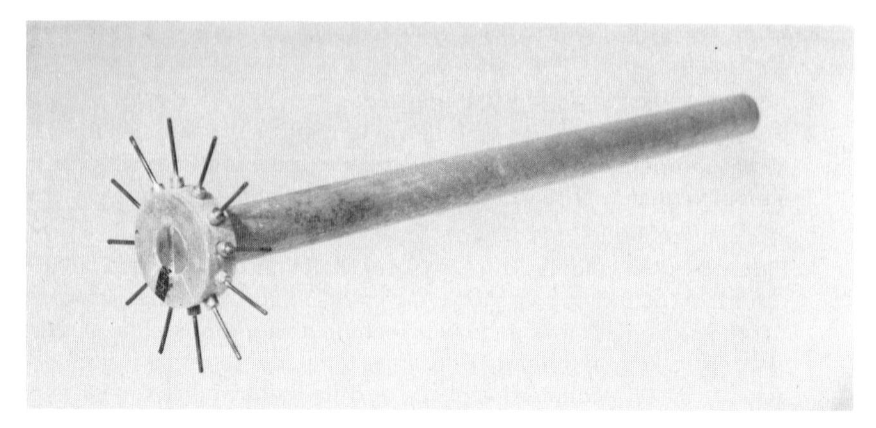

B

Figure 3. A A 12-pin 'wheel-picker' device that speeds up manual picking of colonies. B A 96 pronge spotting comb that can be used manually or automatically for the duplication of cultures in microtitre plates, and inserted into a robot holder for automated high density spotting.

that have been inoculated away from those still to be inoculated on two parts of the bench separated by the individual's working space and the sterilization area. It is ideal to place the inoculated plates directly into an appropriate box for transportation into the 37 °C incubator. The Nunc plates should be placed on a black surface to facilitate visualization of colonies. A note book should be readily available in which to note errors.

It is important to pay attention while picking colonies. Do not pick colonies that are too closely spaced, irregularly shaped colonies that may be two or more colonies fused, or any colonies which differ in appearance. Bear in mind that cosmid clones grow to different sizes so that it is important to pick the very small ones.

It is important to be able to clearly see the first pin on the wheel-picker. Count to 12 while picking. Hold the picker vertically so no pins other than the one intended can touch the agar. Make sure the 12th pin reaches the medium in the last well. It is necessary to tilt the plate slightly with the other hand, to ensure the final pin inoculates a well. At the same time make a scratch on the plastic next to the row just completed. Make sure you see the previous scratch before inoculating the next row.

In order to carry out automatic picking our laboratory uses a robot, of our design (23), that is commercially available through Genetix (Wimborne, Dorset, UK). Follow the protocol provided with the machine for picking colonies. It is essential to pay utmost attention to the density and size of the colonies. Calibrate the machine and carry out a small test run to check its efficiency.

3.2 Duplicating libraries in microtitre plates

The creation of identical sets of the library on microtitre plates from the original set is carried out either manually or automatically using multiwell devices (combs). These are ideally hollow in order to transfer some volume (usually less than 5%) of the microtitre plate culture into the recipient microtitre plate culture medium. Alternatively devices carrying straight metal pins (like the spotting gadget, see *Figure 3B*) or plastic disposable pin gadgets (Genetix). Immerse the replicating device slowly into the source plate and stir gently for about 15 sec to resuspend the cells. Transfer the gadget to the first duplicate plate, and stir four rounds. Transfer the device back to the source and stir four rounds before transferring to the next duplicate plate. Stir four rounds and continue in this way until last duplicate plate is inoculated. Blot off the excess of liquid on a clean stack of Whatman paper, and immerse the device into ethanol. Blot off the excess ethanol on to another stack of Whatman paper. Place the device on a hot plate for 30 sec and then let it cool upside-down for 2–3 min. Take care that the level of ethanol is sufficiently high. Continuously top it up when it evaporates, and occasionally change the entire dish for fresh ethanol. Alternatively, plastic, autoclavable duplicating combs can be used, one per microtitre plate.

3.3 Storage, maintenance, freezing, and defrosting of libraries on microtitre plates

1. Storage guide-lines for cosmids. Cosmids replicate autonomously of the genome of *E. coli*, but due to their high copy number, can impose a significant disadvantage on the host cell. Consequently, the longer the cells are left to metabolize after growth to saturation, the greater the chance for the rearrangements to occur to the cosmid DNA. Bacteria carrying cosmids are in

general cultured for 14–16 h at 37 °C in microtitre plates without shaking or for 12 h with shaking. The plates should then be left at room temperature or at 4 °C for the shortest possible time necessary before freezing. They should not be kept at 4 °C for more than two or three days. Ideally, plates should be defrosted, used, and re-frozen the same day. When plates are left at 4 °C, extra care has to be taken that all plates are properly wrapped, and that an additional wrap is placed over the whole box to prevent evaporation from the wells.

2. Maintenance guide-lines. It is usual to keep at least four copies of each library. The original plates are not used, except to regenerate other copies, if necessary. Separate plates are used for replicating the library and for picking clones. As the content of the wells of the plates used for replication (spotting) becomes slightly contaminated by neighbouring clones during the spotting operation, it is essential, that this copy is never re-replicated, and not be used to pick clones.

3. Freezing. Microtitre plates should be frozen in stacks of six oriented the same way. Each stack of six plates has to be wrapped individually in Saran wrap. Three stacks should be aligned along their longer sides into a 'brick' and taped together. Each brick should be labelled to indicate the library, the duplicate version, and the numbers of the microtitre plates. Each brick has to be packed into a nylon bag which is heat sealed with the labelling clearly visible from the outside.

4. Defrosting. For best results, the microtitre plates have to be immediately unwrapped and laid on the bench singly to defrost. After they are defrosted, if there is too much condensation on the lids, they can be wiped using a clean dry sterile tissue.

3.4 Replicating (spotting) clones using a robotic device

Depending on the design and the degree of automation of the device used to spot cultures on to nylon membranes, it may be necessary to modify *Protocol 8*. This protocol presents the general procedure for spotting colonies at high density on membranes. The ICRF G.A. spotting robot is commercially available from Genetix, Wimborne, Dorset, UK.

Protocol 8. Replicating clones at high density on membranes using the ICRF G.A. robotic device

Equipment and reagents

- 80% ethanol
- 100% ethanol
- 2 × YT broth containing 30 μg/ml kanamycin (see *Protocol 7*)
- 22.3×22.3 cm Hybond N+ membrane (Amersham)
- 3MM Whatman filter paper

Method

1. Wipe down the Perspex surface with 80% ethanol.
2. Replace the used ethanol in the brush box filled completely with fresh 100% ethanol to ensure the replicating pins are washed properly. An example of the 96-pin replicating (spotting) device is shown in *Figure 3B*.
3. Cut 23 cm squares of 3MM paper. Prepare two bottles of 400 ml of 2 × YT containing 30 μg/ml kanamycin. Wearing gloves, soak two Whatman filter paper squares in 2 × YT containing kanamycin in a sterile Nunc dish and let the excess of the medium drip off. Lay the filters down on the Perspex block as square as possible.
4. Put frames on numbered Perspex square blocks. Lay the 22.3 × 22.3 cm Hybond N+ membrane on to the block leaving no more than approximately 1 mm between the edge of the membrane and the frame.
5. Take the frames off and roll the membrane + Whatman flat with either a sterile pipette or a policeman over the filter to remove air bubbles and wrinkles.
6. Turn the robot and computer on and enter the correct program and information.
7. Have the microtitre plates ready to spot in the correct order. Double check the order, and enter it into a file.
8. Vortex each microtitre plate for 1 min at previously determined safe speed immediately prior to spotting.
9. Note any mistakes, unusual observations during the operation (e.g. contamination, unexpected robot actions, etc.) and enter them as errors in the log-book.
10. At the end of the replicating cycle change to the device used for puncturing and enter the punching commands into the computer.
11. Date and label all membranes with a stamp and a pencil.
12. Pick up each membrane by its diagonal corners using Millipore forceps that have been cleaned and wiped in ethanol. Transfer it gently on to a pre-dried plate of 2×YT agar containing 30 μg/ml kanamycin. Place the membrane on to the agar and carefully expel the air slowly towards the diagonal corners. Avoid touching and dragging the forceps over the spotted area! Place the inverted agar plates into the 37° incubator.
13. Re-wrap the microtitre plates and either return them to 40°C or refreeze them.

3.5 Processing membranes

Protocol 9. *In situ* colony lysis, denaturation, and DNA cross-linking of the high density membranes[a]

Equipment and reagents

- Denaturing solution: 0.5 M NaOH, 1.5 M NaCl
- Neutralizing solution: 1 M Tris–HCl pH 7.4, 1.5 M NaCl
- Pre-warmed boiling water-bath
- PROPK solution: 50 mM Tris–HCl pH 8.5, 50 mM EDTA, 100 mM NaCl, 1% Na-lauroyl-sarcosine
- 10 mg/ml proteinase K freshly prepared

Method

1. Examine the membrane and note any abnormalities of colony growth after spotting. Check that the pencil numbers and the punctures made by the robot are clearly visible.
2. Place the membrane on a Whatman 3MM filter paper pre-wetted in denaturing solution for 4 min. Handle the membrane with two Millipore forceps in the diagonal corners.
3. Wipe the water from the lid of the boiling water-bath. Place the membrane on a fresh Whatman filter pre-wet with denaturing solution. Place the filter carrying the membrane on to a glass plate on top of a pipette tip box in the water-bath. The temperature in the water-bath has to be at least 95°C. Leave the bath covered for 4 min.
4. Take out the membrane and place it on a Whatman filter pre-wet in the neutralizing solution and leave for 4 min.
5. Place the membrane on a dry Whatman filter for 1 min.
6. Holding the membrane at two adjacent corners, submerge it into 600 ml of pre-warmed PROPK solution containing 150 μg/ml proteinase K (from the fresh stock solution) at 37°C. Leave in the incubator for 30–50 min.[b] Retrieve the filter holding it as before, and place it on a clean dry Whatman filter.
7. Cover the membrane with an inverted Nunc plate (not the lid) to protect from dust and leave overnight to dry.
8. Next day carry out UV cross-linking. Store the membranes between dry Whatman filters in a dry and safe place.

[a] This protocol is taken from ref. 24.
[b] Avoid shaking or moving the membrane at this time.

3.6 Retrieving single clones

Cosmids are retrieved by scraping a small amount of material from the surface of individual frozen cultures in the multiwell plates and inoculating it into stabs and glycerol stocks. Alternatively, when required for immediate use within the laboratory, frozen stones can be streaked on to agar plates. Special care has to be taken to avoid cross-contamination of the wells by scattering of the frozen material. It is necessary to follow the steps outlined in *Protocol 10*.

Protocol 10. Retrieval of single cosmid clones from frozen microtitre wells

Equipment and reagents

- LB agar for stabs: 200 ml L broth (*see Protocol 6*), 1.5 g agar, autoclave, and aliquot 1.5 ml agar into sterile Nunc cryotubes—flame tubes and caps before replacing caps
- 10 × 10 cm squares of Benchkote each punctured with a single hole size of a single well in the microtitre plate
- Glycerol stocks: aliquot 0.5 ml of the same medium as used for cosmid picking (*Protocol 7*) into Nunc cryotubes—flame tubes and caps before replacing caps
- Ethanol

Method

1. Label stabs and glycerol stock tubes appropriately.
2. Place the microtitre plates on dry ice and maintain them on dry ice during clone retrieval. Place the Benchkote punctured with a single hole over the plate, shiny side up, exposing the well from which culture will be picked.
3. Sterilize the tip of a 2 mm wide screwdriver by wiping it with tissue, immersing it in ethanol, and flaming. Scrape out a small amount of frozen material (pin-head size), and stab it half-way into the agar stab. Use a freshly sterilized screwdriver to scrape off another fragment for a glycerol stock. Discard the protective Benchkote and sterilize screwdrivers after every clone.
4. Re-wrap the microtitre plates back into 'bricks'. Re-seal the bags and return the plates to the − 70°C freezer immediately.
5. Stabs are to be used as soon as possible (within 20 days) and should be kept at room temperature. Glycerol stocks are to be frozen on dry ice and stored at − 70°C.

4. Use of gridded cosmid libraries in hybridization experiments

4.1 Hybridizations with single copy probes

The hybridization of cosmids to single copy probes generally permits the linking of library grid positions to genetic, physical, and cytogenetic maps. On the one hand, unique probes can be derived from cloned genes, cDNA clones, and potentially transcribed sequences. Alternatively, anonymous DNA fragments can be used. This latter category most frequently comprises genetic markers used to establish the distance and order between loci on the genetic map as a result of pedigree analysis of fragment length polymorphism, but may have also been used in the physical mapping experiments. Finally, unique probes can be derived from the ends of inserts carried in other cosmids, YACs, or P1 phages. They are then used to extend existing contigs following the 'chromosome walking' strategy, and potentially interrelating contigs derived from different cloning systems. An example of the hybridization of a genetic marker probe is shown in *Figure 4*. When single copy

Figure 4. An example of a screening of a reference library high density membrane displaying 20 734 cosmids from human chromosome 6 using a single locus probe. The 22 × 22 cm membrane was hybridized to a cDNA probe for the chromosome 6 gene 'DObeta' (38). This probe detected 12 positive cosmid clones. The autoradiogram was kindly supplied by Phillippe Sanseau.

probes contain some fraction of the vector (plasmid) DNA, this can give a non-specific hybridization to a small number (1:1000) of irregularly packaged cosmids (see section 4.2). *Protocol 11* describes the labelling of a single copy probe, and the hybridization and washing conditions using gridded cosmid libraries displayed on high density membrane replicas prepared as described in *Protocols 8* and *9*.

Protocol 11. Hybridization of gridded cosmid libraries to single copy probes

Reagents

- OL: 30 OD U/ml of random sequence hexamer oligonucleotide primers in TE
- TM: 250 mM Tris–HCl pH 8.0, 25 mM $MgCl_2$, 50 mM 2-mercaptoethanol
- Hepes: 1 M Hepes pH 6.6
- LS: 25 parts OL, 25 parts TM, 7 parts Hepes
- dNTP mix: dGTP, dCTP, and dTTP each at 5 mM
- [α-^{32}P]dATP, 3000 Ci/mmol, Amersham
- 5 mg/ml bovine-serum albumin (BSA)
- DNA polymerase large fragment (Klenow) (5 U/μl)
- 0.5 M EDTA
- 10 mg/ml yeast tRNA
- Ethanol
- Dry ice
- 3 M sodium acetate pH 6.0
- TE: 10 mM Tris–HCl pH 8.0, 1 mM EDTA
- [α-^{35}S]dATP, 600 Ci/mmol, Amersham
- Lawrist vector DNA (2–5 ng/μl)
- Church buffer: 0.5 M Na phosphate pH 7.2, 7% SDS, 1 mM EDTA, 0.1 mg/ml yeast tRNA
- 2 × SSC, 0.1% SDS (1 × SSC is 15 mM sodium citrate pH 6.3, 150 mM NaCl)
- 0.1 × SSC, 0.1% SDS

Method

1. To label the probe prepare a 40 μl reaction:
   - 18 μl LS
   - 3 μl of the dNTP mix
   - 5 μl (50 μCi) of [α-^{32}P]dATP
   - 2 μl of BSA
   - 1 μl Klenow

   Denature 20–50 ng of the probe DNA in 11 μl 0.1 × TE by incubating at 98°C for 5 min. Transfer the probe to ice for 1 min. Centrifuge briefly and add the 11 μl of the denatured probe to the above labelling mixture. Leave overnight at room temperature. Add in the following order, mixing after each component: 10 μl EDTA, 3 μl tRNA, 5 μl Na acetate, 145 μl ethanol. Leave for 15 min on dry ice. Centrifuge in a microcentrifuge for 10 min at full speed (14000 r.p.m.). Remove supernatant, briefly air dry the pellet, and dissolve it in 112 μl 0.1 × TE. Count 2 μl (Cherenkov counts) in a scintillation counter.
2. Label 30 ng of Lawrist vector DNA with 25 μCi of [α-^{35}S]dATP with the same basic protocol as in step 1. Purification and counting is not necessary. Alternatively, *E. coli* genomic DNA could be used.[a]

Protocol 11. *Continued*

3. Meanwhile pre-hybridize one or two high density membranes back to back in a plastic bag with 40 ml Church buffer overnight at 65°C.
4. Remove the buffer. Denature the probe by incubation for 5 min at 98°C. Add immediately to fresh Church buffer pre-warmed to 65°C. Use 10^6 Cherenkov c.p.m./ml of Church buffer in a final volume of 15–20 ml per single membrane. At the same time add the denatured vector DNA labelled with ^{35}S. Hybridize overnight at 65°C in a sealed bag.
5. Wash three times with 300 ml of 2 × SSC, 0.1% SDS at room temperature, followed by one wash with 1 litre 0.1 × SSC, 0.1% SDS prewarmed to 68°C, at 68°C for 45 min with gentle rocking (1 litre of buffer).
6. Dry briefly, wrap in Saran wrap, and expose to X-ray film in a cassette with intensifying screen at − 70°C. An 18 h exposure should be enough.[b]

[a] The addition of ^{35}S-labelled vector helps building enough colony background signal to be able to count the coordinates.
[b] If the ^{35}S signal masks a weaker ^{32}P signal, repeat the exposure wrapping Saran wrap plus an additional thicker plastic sheet (such as a Leitz transparent folder), to reduce the ^{35}S signal.

4.2 Hybridizations with complex probes

Complex probes usually have unknown proportions of repetitive sequences shared with the total genomic DNA. This necessitates using protocols that block the hybridization of repetitive sequences, for example by pre-hybridizing the membrane with sheared denatured total genomic DNA, and pre-annealing the denatured probe in solution in the presence of a large excess of such genomic DNA.

These probes can either be localized on longer DNA fragments carried in P1, BAC, or YAC vectors. Alternatively, mammalian DNA from somatic cell
hybrids or radiation fusion hybrids or DNA cut out of the pulsed field electrophoresis gels can be used. These longer fragments are useful for large scale physical mapping, and when used as probes against a cosmid library, provide the most efficient way of integrating long-range physical maps into cosmid maps. When the longer clones are part of an overlap map, groups of cosmids can be assigned to the alternating overlap non-overlap segments between the larger clones defined by the overlap map. In the case of irradiation hybrids, this results in a rapid assignment of large, but incomplete, groups of cosmids to the one to ten megabase intervals between irradiation hybrid segments (25, 26). If the starting point is a YAC overlap map, this can result in a 'cosmid pocket' map, where complete groups of cosmids are assigned to 100–500 kb

intervals (pockets) (27). Cosmids inside each pocket can then easily be turned into a cosmid contig by detecting overlaps using gel-based fingerprinting or restriction enzyme mapping. Complete contigs have in this way been established in small chromosomal regions such as those associated with loci for genetic disease (28, 29). Cosmid pocket maps are starting to appear on the scale of entire human chromosomes (27). If the overlaps between larger fragments are not known, their hybridization to cosmids can be used to obtain such information. Complex probes representing large fragments can be:

- intact large fragments, like YAC clones or genomic restriction fragments, cut out of the pulsed field electrophoresis gel
- mixtures of short DNA fragments derived from the defined large fragments (YACs, hybrids) by means of inter-repeat PCR (25–27, 30)

Complex probes not attached to a single locus or region can also be used. These can be pools of clones or clone-end probes used in multiplex walking strategies. These pools can be from a permuted matrix of rows and columns. This also permits the detection of overlaps in a cosmid library on a large scale (31). Complex populations of mRNA or cDNA libraries from a particular tissue may also be used (32) to help identify cosmids bearing transcribed sequences, and construct integrated transcriptional maps. An example of the hybridization of a complex probe to a high density membrane of a flow sorted human chromosome cosmid library is shown in *Figure 5*. In this experiment, a YAC with an insert greater than 1 Mb has been cut out of a PFG and hybridized to the human chromosome 21 library as a part of the strategy to build the cosmid pocket map of this chromosome (27). Most other complex probes can be hybridized using slight variations of the basic method in *Protocol 12*, steps 5 and 6. Although highly repetitive sequences (such as *Alu*) are usually competed out by efficient protocols (*Protocol 12*), other sources of non-specific hybridization remain. These are, for example, intermediate and low copy repeats, such as O-type repeats, transposons, and other similar types of sequences. Ribosomal DNA bearing cosmids appear positive on all YAC hybridizations due to the homology in the ribosomal sequences between yeast and mammalian genomes. About one in 10^3 cosmids in a Lawrist vector library contain clones which result from irregular packaging events. These contain more than one copy of the vector, and usually retain the fragment between the two 'cos' elements, normally deleted by the packaging process in the majority of clones. If this fragment is present, it can be a source of non-specific hybridization due to its homology to some cloning vectors, such as pYAC4 (D. N. unpublished observation). It is a good idea to hybridize the cosmid library membrane to a vector (e.g. Lawrist4, pBR322, or pYAC4) labelled with ^{35}S in order to identify such irregular cosmids in advance, and ignore them in the subsequent interpretations of the hybridization results.

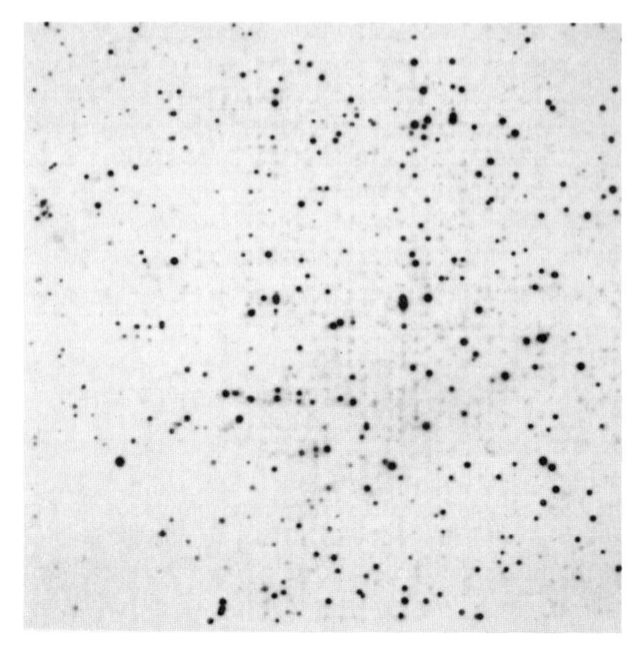

Figure 5. An example of the hybridization of a YAC to the cosmid reference library membrane. A chromosome 21 YAC (3) 767d6ceph (~ 1.4 Mbp) was excised from a pulsed field gel as described in the text and hybridized to the chromosome 21 cosmid library membrane containing 20 734 colonies (in effect seven to eight chromosome equivalents). A total of 404 positive cosmids with differing signal intensities were detected as a part of the effort to construct the cosmid pocket map of this chromosome (27).

Protocol 12. Screening the high density cosmid membranes with YAC probes[a]

Equipment and reagents

- Yeast cells lysed in low melting temperature agarose blocks (34)
- TE: 10 mM Tris–HCl pH 8.0, 1 mM EDTA
- TBE: 10 mM Tris–HCl pH 8.0, 10 mM borate, 10 mM EDTA
- 1 μg/ml ethidium bromide in water
- 4 M NaCl
- Agarase (2 U/μl) (Calbiochem)
- Phenol (distilled and neutralized with Tris–HCl pH 8.0)
- 24 parts chloroform: 1 part isoamyl alcohol
- Ether: water-saturated diethylether
- Absolute ethanol
- 70% ethanol
- Dextrane T-40, 10 mg/ml fresh solution in sterile water (BDH)
- OL: 30 OD U/ml of random sequence hexamer oligonucleotide primers in TE
- TM: 250 mM Tris–HCl pH 8.0, 25 mM $MgCl_2$, 50 mM 2-mercaptoethanol
- Hepes: 1 M Hepes pH 6.6
- LS: 25 parts OL, 25 parts TM, 7 parts Hepes
- dNTP mix: dGTP and dTTP each at 5 mM
- [α-^{32}P]dATP, and dCTP 3000 Ci/mmol, Amersham
- 5 mg/ml bovine serum albumin (BSA)
- DNA polymerase large fragment (Klenow) (5 U/μl)
- 0.5 M EDTA
- 10 mg/ml yeast tRNA
- Ethanol
- Dry ice

- 3 M sodium acetate pH 6.0
- TE: 10 mM Tris–HCl pH 8.0, 1 mM EDTA
- [α-^{35}S]dATP, 1000 Ci/mmole Amersham
- Sheared human placental DNA (Sigma, 10 mg/ml stock)
- 1 M sodium phosphate pH 6.8
- Lawrist vector DNA (2–5 ng/μl)
- Church buffer: 0.5 M Na phosphate pH 7.2, 7% SDS, 1 mM EDTA, 0.1 mg/ml yeast tRNA
- 2 × SSC, 0.1% SDS (1 × SSC is 15 mM sodium citrate pH 6.3, 150 mM NaCl)
- 0.2 × SSC, 0.1% SDS

A. *Preparation of probes*

1. Prepare blocks containing cells with YACs in LMP agarose using standard protocols such as given in ref. 34 (see also Chapter 4). Wash the blocks extensively in 1 × TE.
2. Subject the blocks to PFGE using the CHEF-Mapper from Bio-Rad system at 40 sec for 16 h, followed by 80 sec for 12 h, followed by 110 sec for 12 h in 0.5 × TBE at 5.2 V/cm 14°C. This separates the range between 400–900 kb. It may be necessary to use different switching times for YACs shorter or longer than this.
3. Stain the gel with ethidium bromide at least for 1 h at room temperature. Examine the gel using a long wave UV hand-held lamp (360 nm wavelength). If you see a separate YAC band, as an additional band against the yeast host marker, cut it out. If you don't see a band, photograph the gel on the transilluminator, transfer the DNA to a nylon membrane, hybridize the membrane to ^{32}P-labelled total human DNA, and locate the position of the band hybridizing to human DNA. In parallel to these preparations, run another, identical pulsed field gel. With reference to the autoradiogram of the first PFGE filter, examine the second PFGE under long wave UV (360 nm). (DO NOT EXPOSE IT TO SHORT WAVE UV AT ALL, NOT EVEN FOR A SECOND!!!). If it is possible to estimate the position of the YAC with reference to a yeast chromosome from the hybridization pattern of the first gel, cut out that region of the second gel.
4. Put the excised blocks corresponding to a volume of 300 μl into a microcentrifuge tube. Add 200 μl 1 × TE, and adjust to a final concentration of 100 mM NaCl. The final volume should not exceed 500 μl. Melt the block by incubating for 10 min at 68°C. Transfer the tube to 37°C for 20 min. Add 10 U of agarase and leave overnight at 37°C. The next day perform one extraction with a 500 μl of a 1:1 mixture of phenol and chloroform isoamyl alcohol. Centrifuge for 3 min. Transfer approximately 400 μl of the aqueous phase to a clean microcentrifuge tube avoiding the large white interphase. Add 40 μl 3 M sodium acetate pH 6.0, and 5 μl of freshly prepared solution of dextrane T-40 in water to the aqueous phase. Mix, and add 1 ml of ethanol. Leave overnight at −20°C. Centrifuge in the microcentrifuge at 14000 r.p.m. for 15 min. Wash the pellet with 70% ethanol. Dry it briefly and dissolve in 16 μl

Protocol 12. *Continued*

of 0.1 × TE. Pipette carefully as the dextrane pellet takes some time to dissolve and tends to stick to the plastic tip. If you have pooled more than three blocks dissolve in a larger volume. For each additional block, add another 5 μl 0.1 × TE.

5. Label 8–10 μl of the YAC DNA from the pools of one to three blocks. Prepare the following 40 μl reaction:
   - 18 μl LS
   - 3 μl of the dNTP mix
   - 4 μl (40 μCi) of each [α-^{32}P]dATP and dCTP
   - 2 μl of BSA
   - 1 μl Klenow polymerase

   Denature 20–50 ng of the probe DNA in 11 μl 0.1 × TE by incubating at 98°C for 5 min. Transfer the probe to ice for 1 min. Centrifuge briefly and add the 11 μl of the denatured probe to the above labelling mixture. Leave overnight at room temperature. Add in the following order, mixing after each component: 10 μl EDTA, 3 μl tRNA, 5 μl Na acetate, 145 μl ethanol. Leave for 15 min on dry ice. Centrifuge in a microcentrifuge for 10 min at full speed (14 000 r.p.m.). Remove supernatant, briefly air dry the pellet, and dissolve it in 112 μl 0.1 × TE. Count 2 μl (Cherenkov counts) in a scintillation counter.
6. To pre-anneal the remaining 110 μl of the labelled YAC add 10 μl of sheared placental human DNA (Sigma, 10 mg/ml stock), denature for 10 min at 100°C. Transfer to 65°C and after 10 min, add 17 μl 1 M sodium phosphate pH 6.8. Leave for 2.5–4 h at 68°C.
7. Label 30 ng of Lawrist vector with 25 μCi of [α-^{35}S]dATP using the same basic protocol as in step 5. Purification and counting is not necessary. Alternatively, *E. coli* genomic DNA could be used to prepare a probe to detect the colony pattern.[b]

B. *Hybridization procedure*

1. Pre-hybridize one high density membrane in a bag with 40 ml Church buffer containing an additional 50 μg/ml sheared placental human DNA (Sigma, diluted from 10 mg/ml stock). Incubate overnight at 65°C.
2. Remove the buffer. Add the pre-annealed probe directly to fresh Church buffer pre-warmed to 65°C, with no additional placental DNA. Use 2 × 10^6 Cherenkov counts of probe /ml of the Church hybridization buffer and a final volume of 20–40 ml. Hybridize overnight at 65°C.
3. Pre-wash the filters three times with 300 ml of 2 × SSC, 0.1% SDS at room temperature. Then wash once with 1 litre 0.2 × SSC, 0.1% SDS pre-warmed to 68°C, at 68°C for 45 min.

4. Dry the filter briefly, wrap, in Saran wrap and expose to X-ray film in a cassette containing an intensifying screen at −70°C. A 24–36 h exposure should be optimal.[b]

[a] The original protocol is described in ref. 33.
[b] Usually, there is no need to add ^{35}S-labelled vector DNA, because the YAC insert hybridizations leave enough background to be able to count the coordinates. However, 30 ng of Lawrist vector can be labelled with 25 μCi of [α-^{35}S]dATP as in *Protocol 11*. Denature it and add it to the 40 ml hybridization volume. If the ^{35}S signal masks a weaker ^{32}P signal, repeat the exposure wrapping Saran wrap plus an additional thicker plastic sheet (such as a Leitz transparent folder), to reduce the ^{35}S signal.

4.3 Hybridizations with dispersed probes

Simple but dispersed probes can also be used in fingerprinting by hybridization (1). These probes are devoid of unwanted repetitive sequences, and are designed to hybridize to a large fraction (5–50%) of all cosmids. After a battery of such probes is subsequently hybridized to the same library, fingerprints of cosmids can be expressed as strings of positive or negative hybridizations. Pairwise comparisons of these strings may reveal potentially overlapping cosmids (35, 36). This technique represents a major task for the image scoring (image analysis) and likelihood calculations. It is best to combine it with the cosmid pocket approach to reduce the length of the target contigs (27, 36). Probes which are used for such an approach are short random and anonymous oligonucleotide sequences, polyA-containing and A/T-rich oligonucleotide sequences (37), and simple tandem repeat sequences (microsatellites). *Alu*, LINE, O-type, transposon-like, and MERs sequences can also be used to design the oligonucleotides. Examples of such a hybridization are shown in *Figure 6*. The identification of simple tandem repeats (microsatellites) on mapped cosmids can speed up the generation of flanking sequences, which serve as primers in PCR for the detection of microsatellite length polymorphisms. This can link genetic data with a specific group of cosmids mapping near a locus of interest. Another reason for using short dispersed sequences is to identify cosmids representing particular genomic regions with a known function. These can be alphoid repeats, telomeric repeats, ribosomal units, zinc-finger motifs, and so on.

Protocol 13. Hybridizations using short oligonucleotides[a]

Equipment and reagents

- 1 × SSC: 15 mM sodium citrate pH 6.3, 150 mM NaCl
- Sodium lauroyl sarcosine (30% w/v stock) (NaDodSarc, Sigma)
- 10 × kinase buffer: 500 mM Tris–HCl pH 7.6, 100 mM $MgCl_2$, 5 mM dithiothreitol
- T4 polynucleotide kinase (Biolabs, 10 U/μl)
- 500 mM EDTA
- [γ-^{32}P]ATP, 6000 Ci/mmol (Amersham)
- PEI thin-layer chromatographic plates (CamLab)
- Buffer for PEI: 750 mM potassium phosphate pH 4.0

Protocol 13. *Continued*

Method

1. Label oligonucleotides using 4 μl [γ-^{32}P]ATP per 15 pmol of oligonucleotide in a 10 μl reaction containing 1 μl 10 × kinase buffer and 0.5 μl T4 polynucleotide kinase. Incubate for 40 min at 37°C. Stop the reaction by adding EDTA to a final concentration of 10 mM.
2. Analyse the extent of incorporation by PEI thin-layer chromatography. An incorporation of at least 75% will give a suitable probe.
3. Pre-hybridize the membranes at room temperature for 1 h in 4 × SSC, 7% Na-lauroyl-sarcosine, 1 mM EDTA.
4. Add the terminated reaction mixture directly into hybridization buffer (4 × SSC, 7% Na-lauroyl-sarcosine, 1 mM EDTA) with no further purification. Hybridize overnight. The temperature of the hybridization and wash varies depending on the length and sequence of the oligonucleotide. The best suggestion is to hybridize at 4°C.
5. Wash stepwise for 1 h per wash (in the same buffer). Expose the filter to X-ray film at −70°C to follow the effectiveness of each washing step. The starting wash temperature could be 4°C for oligonucleotides of 7–11 bases, or for longer but A/T-rich sequences. Room temperature may be used for oligonucleotides of 12–15 bases. 37°C may be used for longer and/or C/G-rich oligonucleotides.

[a] This protocol is also given in ref. 24.

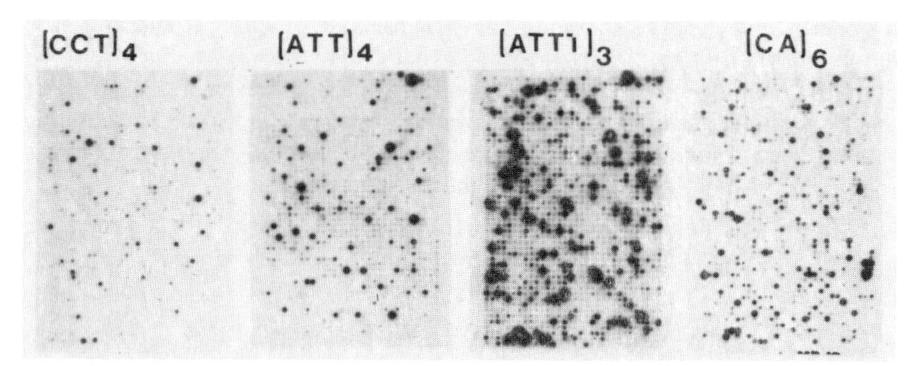

Figure 6. Multiple examples of hybridizations of simple tandem repeat oligonucleotides (sequences indicated on top) in the form of 12-mers to a cosmid high density membrane. 11 × 7 cm membranes containing 1536 cosmids from the chromosome X cosmid library (12) were used. The frequencies of positive cosmids varied between 3–40%, and show a wide range of signal intensities.

5. Concluding remarks

In this chapter we have presented a set of techniques which were developed in our group to meet the needs of the parallel approach to genome mapping. The use of chromosome-specific libraries decreases the amount of experimental work in mapping of single chromosomes. The storage of the library in microtitre plates and the display on membranes at high density allow rapid and parallel generation of mapping information on the entire set of cosmids representing the chromosomal DNA. It also enables direct cross-referencing of data generated by many different groups on the same cloned material. The hybridization techniques allow easy linking of the cosmid contigs to overlap maps generated in other cloning systems such as YACs. This directly places groups of cosmids on the long-range physical map of the chromosome defined by the maps of sized, overlapping YACs (the 'cosmid pocket' approach). This also has the advantage of partially ordering the entire chromosome library in cosmids and reducing the number of cosmids that need to be compared for obtaining fully ordered overlapping cosmids, to such that can easily be handled on single electrophoresis gels. Such 'pockets' represented as pools of unordered but potentially overlapping cosmids, are also a good starting material for the detection of transcripts using exon trapping, cDNA fishing, or related approaches. In this way, a transcriptional map is directly linked to overlap and physical maps. Pools of cosmids are also good FISH probes, which can be used in the mapping of deletion syndromes and other diagnostic purposes, such as prenatal detection of aneuploidies or monitoring of the presence of cancer cells with a specific chromosomal rearrangement on interphase nuclei. Hybridizations of polymorphic markers to gridded cosmids can then link all this information to the genetic map, and through pedigree analysis, lead to the functional (morbid) map in regions associated with genetic diseases and disorders. The result is a faster and easier generation of a genome map that integrates many different types of information.

Acknowledgements

We thank Pillipe Sanseau for the autoradiogram of an unpublished hybridization, Carola Burgtorf and Guenther Zehetner for the critical reading of the manuscript, and Simon Monard and Bryan Young for the protocols for flow sorting. The team at the Genome Analysis Department of ICRF, or closely collaborating with it, all helped in the development and refinement of various protocols described. Among them, we are in particular thankful to: Guenther Zehetner, Joerg Hoheisel, Alister Craig, Lisa Gellen, Mark Ross, Radoje Drmanac, Elmar Maier, Sarah Baxendale, Sebastian Maier-Ewert, Ali Ahmadi, Jon Curtis, Annemarie Poustka, and Anthony Monaco.

References

1. Lehrach, H., Drmanac, R., Hoheisel, J. D., Larin, Z., Lennon, G., Monaco, A. P., *et al.* (1990). In *Genome analysis* (ed. K. E. Davies and S. Tilghman), pp. 39–81. Cold Spring Harbor Laboratory Press, Cold Spring Harbor, NY.
2. Burke, D. T., Carle, G. F., and Olson, M. V. (1987). *Science*, **236**, 806.
3. Chumakov, I., Rigault, P., Guillou, S., Ougen, P., Billaut, A., Guascono, G., *et al.* (1992). *Nature*, **359**, 380.
4. Poustka, A. and Lehrach, H. (1985). In *DNA cloning: a practical approach* (ed. D. Glover), pp. 43–57. IRL Press, Oxford.
5. Sternberg, N. (1990). *Proc. Natl Acad. Sci. USA*, **87**, 103.
6. Hosoda, Nishimura, S., Uchida, S., and Ohki, M. (1990). *Nucleic Acids Res.*, **18**, 3863.
7. Korn, B., Sedlacek, Z., Manca, A., Kioschis, P., Konecki, D., Lehrach, H., *et al.* (1992). *Hum. Mol. Gene.*, **1**, 235.
8. Buckler, A. J., Chang, D. D., Brook, J. D., Harber, D. A., Sharp, P. A., and Housman, D. E. (1991). *Proc. Natl Acad. Sci. USA*, **88**, 4005.
9. Lovett, M., Kere, J., and Hinton, L. M. (1991). *Proc. Natl Acad. Sci. USA*, **88**, 9628.
10. Parimoo, S., Patanjali, S. R., Shukla, H., Chaplin, D. D., and Weissman, S. M. (1991). *Proc. Natl Acad. Sci. USA*, **88**, 9623.
11. Nizetic, D., Figueroa, F., Dembic, Z., Nevo, E., and Klein, J. (1987). *Proc. Natl Acad. Sci. USA*, **84**, 5828.
12. Nizetic, D., Zehetner, G., Monaco, A. P., Gellen, L., Young, B. D., and Lehrach, H. (1991). *Proc. Natl Acad. Sci. USA*, **88**, 3233.
13. Zehetner, G. and Lehrach, H. (1994). *Nature*, **367**, 489.
14. Nizetic, D., Monard, S., Cotter, F., Young, B. D., and Lehrach, H. (1994). *Mamm. Genome*, **5**, 801.
15. Potier, M. C., Kuo, W. L., Dutriaux, A., Gray, J., and Goedert, M. (1992). *Genomics*, **14**, 481.
16. Gray, J. (ed.) (1989) *Flow cytogenetic. Analytical cytology series.* AP Harcourt B.J.
17. de Jong, P. J., Yokobata, K., Chen, C., Lohman, F., Pederson, L., McNinch, J. *et al.* (1989). *Cytogenet. Cell Genet.*, **51**, 985.
18. Deaven, L. L., Van Dilla, M. A., Bartholdi, M. F., Carrano, A. V., Cram, L. S., Fuscoe, J. C., *et al.* (1986). *Cold Spring Harbor Symp. Quant. Biol.*, **51**, 159.
19. Gibson, T. J., Coulson, A. R., Sulston, J. E., and Little, P. F. (1987). *Gene*, **53**, 275.
20. Poustka, A., Rackwitz, H. R., Frischauf, A. M., Hohn, B., and Lehrach, H. (1984). *Proc. Natl Acad. Sci. USA*, **81**, 4129.
21. Murray, N. E., Brammar, W. J., and Murray, K. (1977). *Mol. Gen. Genet.*, **150**, 53.
22. Hoheisel, J. D., Nizetic, D., and Lehrach, H. (1989). *Nucleic Acids Res.*, **17**, 9571.
23. Maier-Ewert, S., Maier, E., Ahmadi, A., Curtis, J., and Lehrach, H. (1993). *Nature*, **361**, 375.
24. Nizetic, D., Drmanac, R., and Lehrach, H. (1991). *Nucleic Acids Res.*, **19**, 182.

25. Monaco, A. P., Lam, V. M. S., Zehetner, G., Lennon, G. G., Douglas, C., Nizetic, D., *et al.* (1991). *Nucleic Acids Res.*, **19**, 3315.
26. Kumlien, J., Labella, T., Zehetner, G., Vatcheva, R., Nizetic, D., and Lehrach, H. (1994). *Mamm. Genome*, **5**, 365.
27. Nizetic, D., Gellen, L., Hamvas, R., Mott, R., Grigoriev, A., Vatcheva, R., *et al.* (1994). *Hum. Mol. Genet.*, **5**, 759.
28. Zuo, J., Robbins, C., Baharloo, S., Cox, D. R., and Myers, R. M. (1993). *Hum. Mol. Genet.*, **2**, 889.
29. Baxendale, S., MacDonald, M. E., Mott, R., Francis, F., Lin, C., Kirby, S. F., *et al.* (1993). *Nature Genet.*, **4**, 181.
30. Cole, C. G., Goodfellow, P. N., Bobrow, M., and Bentley, D. R. (1991). *Genomics*, **10**, 816.
31. Evans, G. A. and Lewis, K. A. (1989). *Proc. Natl Acad. Sci. USA*, **86**, 5030.
32. Hochgeschwender, U., Sutcliffe, G. J., and Brennan, M. B. (1989). *Proc. Natl Acad. Sci. USA*, **86**, 8482.
33. Baxendale, S., Bates, G. P., MacDonald, M. E., Gusella, J. F. and Lehrach, H. (1991). *Nucleic Acids Res.*, **19**, 6651.
34. Larin, Z., Monaco, A. P., and Lehrach, H. (1991). *Proc. Natl Acad. Sci. USA*, **88**, 4123.
35. Craig, A. G., Nizetic, D., Hoheisel, J. D., Zehetner, G., and Lehrach, H. (1990). *Nucleic Acids Res.*, **18**, 2653.
36. Hoheisel, J. D., Maier, E., Mott, R., McCarthy, L., Grigoriev, A. V., Schalkwyk, L. C., *et al.* (1993). *Cell*, **73**, 109.
37. Drmanac, R., Nizetic, D., Lennon, G. G., Beitverda, A., and Lehrach, H. (1991). *Nucleic Acids Res.*, **19**, 5839.
38. Tonnelle, C., DeMars, R., and Long, E. O. (1985). *EMBO J.*, **4**, 2839.

3

Library construction in P1 phage vectors

N. STERNBERG

1. Introduction

The P1 cloning system was developed as an alternative to YAC and cosmid technology for the cloning of high molecular weight DNA fragments. It permits the recovery of inserts in *E. coli* that are up to 95 kb in size (more than twice the size of inserts that can be recovered in cosmids) with an efficiency that is intermediate between λ cosmid and YAC cloning; greater than 10^5 clones can be generated using the P1 system with 1–2 μg of vector arms and 2–4 μg of *Sau*3AI digested, size selected genomic DNA insert (1). While the size of cloned inserts recovered with the P1 system are significantly smaller than are those recovered with YACs, the former offers several advantages: libraries representative of multiple copies of the genome are more easily produced, larger quantities of a particular cloned DNA are more easily isolated, the cloning process may be more faithful (2, 3), and physical dissection of cloned inserts may be more easily accomplished. Indeed, the ability to analyse cloned DNA before and after attempting amplification with the P1 lytic replicon permits one to assess whether amplification of the insert by a high copy number replicon can indeed occur, and if so, whether DNA integrity is maintained.

2. Background information

2.1 Principles of P1 packaging

Bacteriophage P1 packages its DNA by a processive headful mechanism. The substrate for this reaction is a concatemer that is generated by rolling circle replication and consists of tandemly repeated units of the phage genome arranged in a head-to-tail configuration. Packaging is initiated when a unique 162 bp P1 *pac* site is recognized and cleaved by phage-encoded pacase proteins (4). One of the cleaved *pac* ends is brought into an empty phage prohead and the head is then filled with that DNA. Once the head is full the packaged DNA is cleaved away from the rest of the concatemer by a ‘headful’ cutting reaction that appears not to recognize any specific DNA sequence. A second round of packaging is initiated from the unpackaged end produced by the headful cut. These events produce a processive series of DNA headfuls starting from a single *pac* site on DNA. A P1 headful is about 110 kb. The 162 bp *pac* site contains four hexanucleotide sequences (5′ TGATCA/G) at one end, and three at the other, and a 90 bp region between them. *Pac* cleavage is localized to about a turn of the DNA helix near the centre of the 90 bp region. Each hexanucleotide sequence contains a DNA adenine methylation (*dam*) sequence (5′ GATC) and *pac* must be methylated for it to be cleaved (5). It is our experience that vectors containing *pac* are more readily cleaved by the stage I extract if the plasmid is prepared without an amplification step that inhibits cellular protein synthesis. Presumably, plasmid DNA is undermethylated under these conditions.

2.2 The p*Ad*10*sac*BII cloning vector

This vector is divided into two domains by flanking *lox*P recombination sites (*Figure 1*). The '*amp*^{r}' domain contains the multicopy *col*E1 (pBR322)

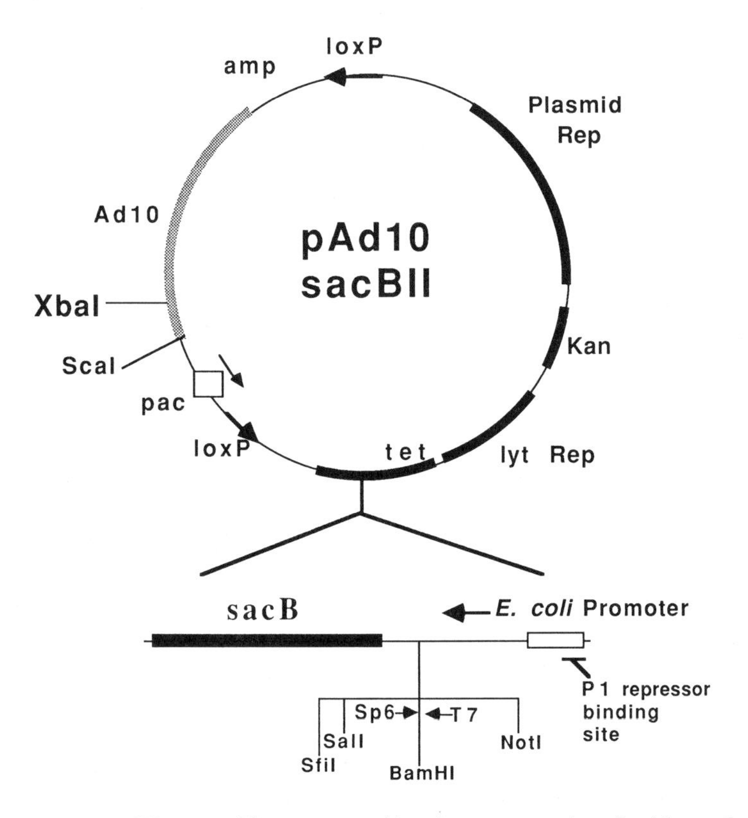

Figure 1. p*Ad*10*sac*BII vector. The vector and its elements are described in section 2.2.

replicon, the P1 *pac* site oriented so as to direct packaging counterclockwise, and an 11 kb *Bam*HI–*Sca*I 'stuffer' fragment from adenovirus DNA cloned into the *Sca*I site of the amp^r (*bla*) gene so as to retain the unique *Sca*I site in the vector at a location just clockwise from *pac*. The kan^r domain contains the kan^r gene from transposon Tn903, the tet^r gene from plasmid pBR322, the unit copy P1 plasmid replicon and partition system (6), and a P1 lytic replicon whose activity is regulated by the *lac* operon promoter (7). The latter is inactive in a host containing a $lacI^q$ repressor but can be activated by adding the *lac* inducer IPTG to the growth media. The tet^r gene contains unique *Bam*HI and *Sal*I restriction sites into which foreign DNA fragments were cloned in the precursor to p*Ad*10*sac*BII, p*Ad*10 (8). In p*Ad*10*sac*BII, the *sac*B gene has been cloned between the *Bam*HI and *Sal*I sites of p*Ad*10, destroying these sites. This gene encodes an enzyme that converts sucrose to levan which accumulates in the periplasmic space of cells and results in cell death (9, 10). Thus, cells expressing *sac*B die in media containing greater than 2% sucrose but grow in media lacking the sugar. To take advantage of this property in a positive selection scheme for cloning, an *E. coli* promoter was placed upstream of *sac*B and a *Bam*HI cloning site was placed between these two elements. The *Bam*HI site was in turn flanked by:

- T7 and SP6 promoters for subsequent riboprobe analysis of the ends of any cloned DNA (11)
- rare cutting *Sa*lI, *Sfi*I, and *Not*I restriction sites that permit the cloned DNA to be easily recovered

Finally, a P1 C1 repressor binding site (operator) was positioned so as to overlap the *E. coli* promoter. Under these circumstances transcription of the plasmid borne *sac*B gene is blocked in cells that express P1 C1 (12) and those cells grow better in the absence of sucrose. This helps minimize production of plasmid mutations that inactivate *sac*B gene expression.

2.3 Cloning rationale

The complexity of the P1 cloning vector is designed to permit efficient recovery of unrearranged, insert DNA. In the scheme outlined in *Figure 2* the vector DNA is digested at unique *Sca*I and *Bam*HI restriction sites to generate two vector 'arms'. The short 4.3 kb arm contains *pac, lox*P, and *sac*B, without its promoter, and the long 26 kb arm contains the P1 lytic replicon, the kan^r gene, the P1 plasmid replicon, *lox*P, and the adenovirus stuffer fragment. The ends of the two arms are treated with phosphatase to prevent ligation to each other and the arms are ligated at their *Bam*HI ends to DNA fragments generated by the partial *Sau*3AI digestion of genome DNA. The appropriate ligated product will have the foreign DNA fragment sandwiched between the short and long vector arms such that the two *lox*P sites are oriented in the same direction.

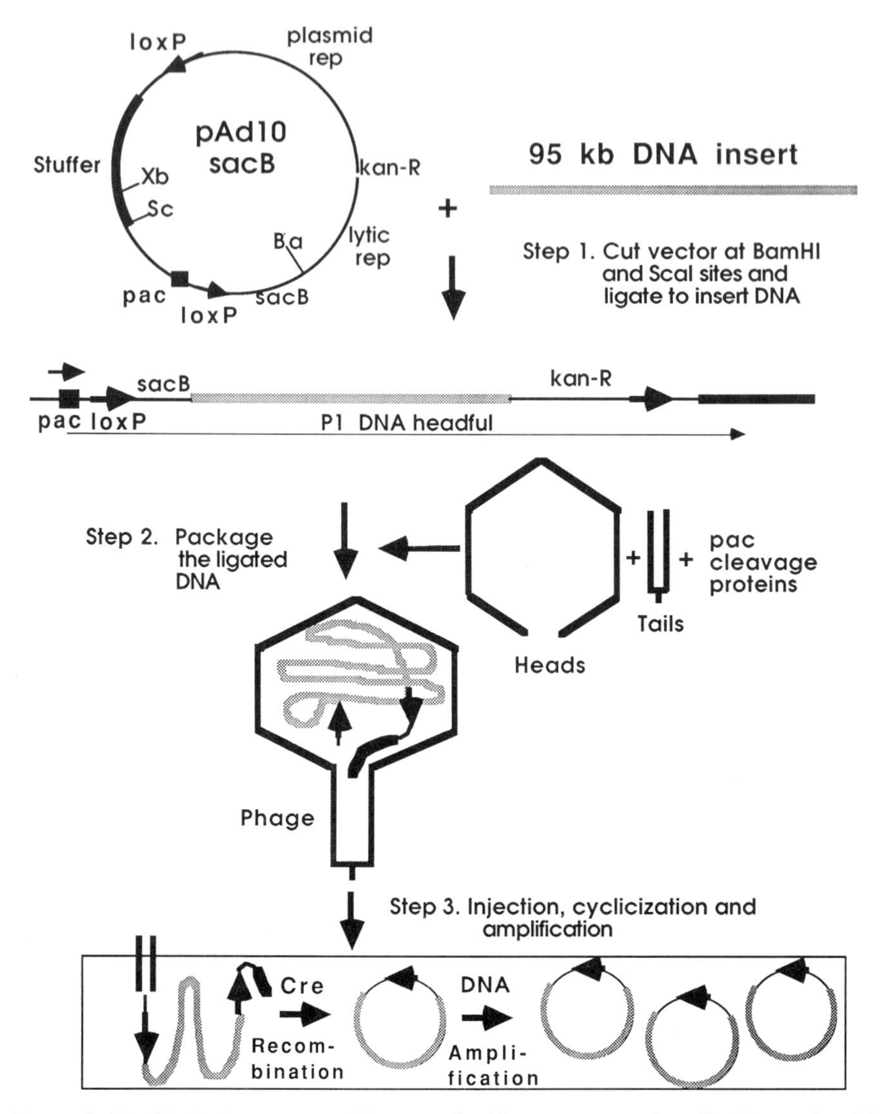

Figure 2. The P1 cloning process. The steps in this process are described in section 2.3.

In the next step of the cloning process the ligated DNA is cleaved by incubating it with an extract that contains the P1 pac cleavage proteins (pacase or stage I extract). This cleaved DNA is then incubated with a second P1 extract (the stage II extract) that contains phage heads and tails. During that incubation the DNA is packaged unidirectionally from its pac cleaved end into the empty phage head until the head is full. At this point the DNA inside of the head is separated from that outside by a headful cutting mechanism and the resulting filled head is converted to an infectious particle by the addition of phage tails.

In the final step of the cloning process the packaged DNA is recovered and amplified. The first step in this process is the cyclicization of the packaged DNA after it is injected into a bacterial host. To achieve this end the packaged insert DNA must be flanked by *lox*P sites present in the vector arms and the recipient cell must express the recombinase (Cre) that promotes recombination between these sites. The host used to recover cloned DNA (NS3529) is also *recA*$^-$, *lacI*^{q}, *mcrABC*, and *mrr*. The *recA* mutation inactivates the homologous recombination system of the host preventing rearrangement of homologous sequences that might be present in the insert DNA. The *lacI*^{q} mutation inhibits activation of the P1 lytic replicon ensuring the infected DNA is maintained in the cell initially at single copy by the P1 plasmid replicon. This minimizes the possibility of rearrangement due to over-replication of the insert. Finally, the *mcr* and *mrr* mutations are employed to inhibit the selective loss of insert DNA that is rich in Gp^{Me}C or possibly Ap^{Me}C (13, 14). For insert DNA in the 70–90 kb size range, results with the P1 system indicate that a functional *mcr* system inhibits cloning about 35-fold (8).

2.4 Packaging small DNA inserts in P1 heads

The results shown in *Table 1* indicate that uncut p*Ad*10*sac*BII vector DNA is packaged with nearly the same efficiency as is vector DNA that has been cut with *Bam*HI and then ligated into a long concatemer. This result is surprising since the vector is only 30 kb in size, not nearly big enough to fill the normal 110 kb P1 head or even the rarer small 47 kb P1 head. We have further examined these results by analysing the equilibrium density of the phage produced when uncut vector is packaged *in vitro*. In CsCl equilibrium gradients these phage have the same density as P1 plaque-forming phage particles; namely, they contain a headful (110 kb) of DNA (N. Sternberg, unpublished data). These results suggest that the P1 head can package multiple copies of less than headful DNA in order to fill its head. Moreover, it explains why the P1 system will recover inserts ranging in size from several kilobase pairs up to 95 kb rather than simply the largest inserts in the fragment pool (85–95 kb) that will fill the phage head and still be flanked by vector *lox*P sites. These observations accentuate the need to produce a 75–95 kb size fractionated population of DNA fragments in order to exclusively clone DNA in this size range. In other words, the system *will not* select the largest clonable fractions from a broad size range of fragments.

3. Preparation of vector and insert DNA

3.1 Preparation of p*Ad*10*sac*BII vector DNA

In this section we describe how we prepare and evaluate the p*Ad*10*sac*BII vector DNA. Greater care must be taken in preparing this plasmid than most because it has a tendency to generate deletions that inactivate *sac*B and would

Table 1. DNA packaging with p*Ad*10*sac*BII[a]

Vector (p*Ad*10*sac*BII)	Insert	T4 DNA ligase	NS3529 Kan(1)	Kan+suc(2)	(2)/(1) × 100
uncut	none	–	1800	1	<0.1
ScaI cut	none	–	1450	4	<0.3
BamH1 cut	none	+	4800	65	1–2
Sca1-BamH1 (CIP)	none	–	50	0	<3
Sca1-BamH1 (CIP)	none	+	320	5	<2
Sca1-BamH1 (CIP)	mouse	+	620	440	70

[a] Typical cloning and packaging results with p*Ad*10*sac*BII. All of these results are generated by assaying 15 μl of a 180 μl packaging reaction. The packaging reaction employs 3 μl of a 10 μl ligation reaction similar to that described in *Protocol 5*.

thus compromise subsequent selection schemes. These deletions arise because cells expressing *sac*B grow more slowly than those that do not, even in the absence of sucrose. To minimize these deletions, one should prepare the plasmid in an *E. coli* strain that constitutively produces P1 Cl repressor. This repressor binds to the P1 operator site in the promoter that is used for *sac*B expression and minimizes transcription of the gene. The *E. coli* strain used is NS3622=DH5α*recA*(λ*imm*λLP1)(λ*imm*21-P1:7Δ5b)(p*Ad*10*sac*BII).

The *imm*21 prophage contains the P1*c*1 gene and the *imm*λ prophage contains a *lacI*^{q} gene.

Protocol 1. Preparation of p*Ad*10*sac*BII vector DNA

Equipment and reagents

- 50 mM Tris–HCl pH 8.0, 25% sucrose
- TE: 10 mM Tris–HCl pH 8.0, 1 mm EDTA
- 5 mg/ml lysozyme in 0.25 M Tris–HCl pH 8.0 (freshly prepared)
- 0.25 M EDTA pH 8.0
- Lysis buffer: 50 mM Tris–HCl pH 8.0, 62 mM EDTA, 2% SDS
- 5 M NaCl
- 50 ml screw-cap Oak Ridge tubes
- SS34 Sorvall rotor, RC-5C Sorvall centrifuge or their equivalent
- Biological grade CsCl (Gallard–Schlessinger 61250)
- Ethidium bromide solution (10 mg/ml)
- Sorvall type T-1270 rotor, Sorvall ultracentrifuge (e.g. type RC-70) or their equivalent
- Phenol redistilled (IBIOS164) saturated with Tris–HCl pH 8.0
- Chloroform/isoamyl alcohol (24:1)
- Isopropanol
- Dialysis tubing (Spectrum No. 132676, MWCO 12–14 000)
- *E. coli* strain NS3622 (see section 3.1 for genotype)
- *E. coli* strain NS3529 (see below for genotype)
- Restriction endonuclease *Spe*I, New England Biolabs (NEB#1335)
- LB agar plates (15)
- LB–kanamycin (25 μg/ml) agar plates (15)
- LB–kanamycin (25 μg/ml) – sucrose (5%) agar plates (15)
- LB broth (15)
- LB broth containing 25 μg/ml kanamycin (15)
- Kanamycin (2.5 mg/ml) in sterile water

Method

1. Transfer each of five colonies of NS3622 from LB–kanamycin agar plates into flasks containing 20 ml LB containing 25 μg/ml kanamycin and culture in a shaking incubator at 37°C until the cells reach a density of 1×10^8/ml ($OD_{590} = 0.3$). Store the cultures for up to several days at 4°C. Remove a 5 ml aliquot from each culture and prepare plasmid DNA by the alkaline lysis procedure (see *Protocol 5*). Resuspend the DNA in 10 μl TE and use 1 μl to transfect calcium competent NS3529 (16, 17). Add 1 ml of LB to the transfected cells and culture them in a shaking incubator for 30 min at 30°C. Spread aliquots on LB–kanamycin agar and LB–kanamycin–sucrose agar plates. Incubate the plates overnight at 37°C and determine which DNA produces the highest LB–kan/LB–kan–sucrose ratio. That ratio should at least be 1000. The culture that was used to produce that DNA will be used in all subsequent steps in this protocol.[a]
2. Dilute the culture of choice into 1 litre of LB containing 25 μg/ml kanamycin and grow until the culture is saturated, usually 12–14 h at 30°C. Pellet the cells by centrifugation (use 350 ml plastic bottles in a GSA rotor at 6000 r.p.m., 4°C in a Sorvall RC-5C centrifuge or equivalent conditions in other systems). Pour off the supernatant, freeze the pellet in a dry ice–ethanol bath, and thaw the pellet at room temperature.
3. Resuspend the pellet in 7 ml cold 50 mM Tris–HCl pH 8.0, 25% sucrose. Transfer to a 50 ml screw-cap Oak Ridge tube and add 1.5 ml of lysozyme (5 mg/ml in 0.25 M Tris–HCl pH 8.0). Keep the cells at 4°C for 15 min.
4. Add 2.8 ml of 0.25 M EDTA pH 8.0 and leave the cells for an additional 5 min at 4°C.
5. Add 11.3 ml of lysis solution and allow the cells to lyse for 5 min at room temperature. Add 5.7 ml of 5 M NaCl and incubate the lysate for 5 min at room temperature, and then 1 h at 4°C.
6. Centrifuge the extract in a Sorvall SS34 rotor at 18 000 r.p.m. for 60 min (or equivalent conditions in an alternative centrifuge) at 4°C. Remove the supernatant and precipitate nucleic acids with 0.54 vol. of cold isopropanol at −20°C for 1 h. Leave at −20°C for 1 h.
7. Pellet the precipitate by centrifugation (SS34 rotor at 10 000 r.p.m., or equivalent, for 10 min). Discard the supernatant and resuspend the pellet in 9 ml of 30 mM NaCl, 10 mM Tris–HCl pH 8.0, 1 mM EDTA.
8. Add 2.7 g of biological grade CsCl and shake gently to dissolve. Add 1 ml of a 10 mg/ml solution of ethidium bromide. Centrifuge the solution in an SS34 rotor (or equivalent) at 8000 r.p.m. for 10 min and discard the pellet. Add an additional 7.4 g CsCl, dissolve, and centrifuge

Protocol 1. *Continued*

the solution at 35 000 r.p.m. in a Sorvall T-1270 rotor (or equivalent) for 42 h.

9. Remove the plasmid band from the side of the gradient with a needle and extract the ethidium bromide with four to five equal volumes of CsCl-saturated isopropanol. After several hours of dialysis against 100 × volume TE at 4°C to remove CsCl, extract the DNA with an equal volume phenol. Re-extract the aqueous phase with an equal volume of chloroform/isoamyl alcohol. Precipitate DNA from the aqueous phase by adding 0.1 vol. of 3 M sodium acetate pH 5.2 and 2 vol. of ethanol. Leave for 1 h at −20°C. Pellet the DNA precipitate by centrifugation in an SS34 rotor (or equivalent) at 12 000 r.p.m. for 10 min and resuspend the pellet in TE buffer. This procedure usually yields about 400–600 μg of vector DNA.
10. To assess the quality of *sac*B vector DNA it should be tested in two ways. First, package it *in vitro* (see below) to determine both the efficiency by which kanamycin-resistant transformants have been generated, and the ratio of kanamycin-resistant transformants to kanamycin–sucrose-resistant transformants. Expected values are shown in *Table 1*. Secondly, 100–200 ng of DNA should be digested with 10 U of restriction enzyme *Spe*I and analysed by standard argarose gel electrophoresis (18). Digestion with this enzyme produces two large fragments (20 kb and 9 kb) and a 1.7 kb fragment that contains the *sac*B gene.

[a] The purpose of this procedure is to select the culture with the fewest number of cells containing vector DNA with an inactive *sac*B gene. Those cells are generated when the *sac*B gene in the vector undergoes a spontaneous mutation or deletion, and are selected for in culture because cells with an inactive *sac*B gene grow faster than those with an active gene. Since vector DNA with an inactive *sac*B gene will grow on LB–kan–sucrose plates, it will compromise the positive selection scheme used for cloning inserts, and needs to be minimized.

3.2 Preparation of genomic DNA fragments for cloning

To clone DNA fragments in the size range 70–100 kb in the P1 system, one must start with high molecular weight genomic DNA that is greater than 300 kb in size and readily digestible by the restriction enzyme *Sau*3AI. In this section we provide a guide for the preparation of such DNA from cultured mammalian cells. We also describe methods for the partial digestion of that DNA by *Sau*3AI in order to clone it into the *Bam*HI site of the pAd10*sac*BII vector. Size fractionations of the digestion products is carried out on sucrose gradients, and DNA is concentrated from the sucrose gradient fractions (*Protocol 2*).

Protocol 2. Preparation of genomic DNA fragments for P1 cloning

Equipment and reagents

- Phosphate-buffered saline: 2.7 mM KCl, 1 mM KH_2PO_4, 8.1 mM Na_2HPO_4, 140 mM NaCl, pH 7.4
- Proteinase K (Boehringer Mannheim 745723) 10 mg/ml
- Bovine serum albumin (BSA), 10 mg/ml (NEB, nuclease-free)
- Phenylmethylsulfonyl fluoride (PMSF, BRL 5521UB)
- 0.1 M $MgCl_2$
- 0.2 M EDTA pH 8.0
- 0.5 × TBE: 45 mM Tris–borate, 1 mM EDTA pH 8.0
- 10 × stop solution: 0.3% bromophenol blue, 0.3% xylene cyanol, 50% glycerol, 30 mM EDTA, 0.2% SDS
- Agarose (BRL, ultrapure grade)
- PC750 FIGE apparatus (Hoefer Scientific)
- DNA markers: yeast chromosome plugs (Bio-Rad 71235), T5 DNA (Sigma P8010), lambda DNA monocut mix (NEB # 301–95), and λ *Hin*dIII digest (NEB # 301–25)
- *Sau*3AI restriction endonuclease (NEB # 169S)
- TE: 10 mM Tris–HCl pH 7.5, 1 mM EDTA
- 10 × magnesium-free *Sau*3AI restriction buffer: 0.1 M Tris–HCl pH 7.5, 1 M NaCl
- Bovine serum albumin (BSA) 3 mg/ml
- 10% and 40% (w/v) sucrose (ultra pure BRL 5503UA) in 20 mM Tris–HCl pH 8.0, 0.8 M NaCl, 10 mM EDTA
- SW28 Beckman rotor (or equivalent)
- RT6000 Sorvall rotor (or equivalent)
- Sucrose gradient maker (Hoefer Scientific)
- Type VS dialysis filters (Millipore, 0.025 μm)

A. *Preparation of genomic DNA*

1. Transfer about 10^8 mammalian cells to a 50 ml conical centrifuge tube. Pellet the cells by centrifugation for 10 min at 1500 r.p.m. at 4°C in an RT6000 Sorvall rotor (or equivalent). Wash the pellet with 2 ml phosphate-buffered saline.
2. Resuspend the cell pellet in 2 ml 50 mM Tris–HCl pH 7.5, 0.1 M NaCl, 0.15 M EDTA, and then add 250 μl 10% SDS and 250 μl 10 mg/ml proteinase K. Incubate at 60°C for 4 h. To prevent shearing of the viscous DNA avoid vigorous mixing or pipetting.
3. Transfer the DNA lysate to a sterile dialysis bag and dialyse against 1 litre of TE containing 0.1 mM phenylmethylsulfonyl fluoride (PMSF) to inactivate the proteinase K for 12 h at 4°C. Repeat the dialysis step four times using TE containing PMSF and then twice using TE without PMSF. This procedure typically yields 2–3 ml of DNA at 100–200 μg/ml.[a]

B. *Partial digestion of HMW genomic DNA*[b]

1. Incubate 100 μl of genomic DNA from the previous step with 12 μl of magnesium-free 10 × *Sau*3AI restriction buffer, 4 μl BSA (3 mg/ml), and 10 U *Sau*3AI for 6 h at 4°C.[c] Also, prepare an equivalent 10 μl reaction without *Sau*3AI. The latter constitutes the no enzyme control.
2. Remove a 10 μl aliquot from the *Sau*3AI sample, incubate it at 70°C for 10 min, and save it at 4°C as a magnesium-free control. Place the remainder of the DNA (110 μl) at 30°C for 1 min, add 12 μl 0.1 M $MgCl_2$, and gently mix it into the DNA solution. Remove 10 μl aliquots at 2 min

Protocol 2. *Continued*

intervals using wide bore tips. Place the aliquots into tubes containing 2 μl 0.2 M EDTA pH 8.0. Heat the tubes immediately to 70°C for 10 min, and then place them on ice. Incubate the no enzyme control tube at 30°C for 20 min and then treat it like the other 30°C incubated samples. Add 2 μl of the 10 × stop solution to each of the samples and analyse the DNA by field inversion gel electrophoresis (FIGE) (step 3). If digestion of the genomic DNA during the 20 min incubation period is either too extensive or not extensive enough repeat the experiment with either less or more enzyme.

3. Fractionate all of the samples generated in step 2 on gels using conditions that resolve DNA in the 20–200 kb range. Load samples into the wells of a 1% agarose gel in 0.5 × TBE with a wide bore tip.[d] Subject the gel to electrophoresis at 100 V for 1 h without a switching regimen. Next subject the gel to electrophoresis in 0.5 × TBE at 160 V with a switching regimen of 0.6 sec forward, 0.2 sec backward, at a ramp factor of 20 for 6 h at room temperature using a PC750 FIGE apparatus. As DNA markers in this analysis use yeast chromosomes in agar plugs (smallest chromosome 225 kb), T4 DNA (165 kb), T5 DNA (122 kb), a T5 *Sac*I, *Xho*I digest (102 kb, 78 kb, 25 kb), λ monocut mix (1.5–48 kb), and a λ *Hin*dIII digest (2–23 kb). At the end of the run, stain the gel by soaking for 1 h in 0.5 × TBE containing 20 μg/ml ethidium bromide. Destain in 0.5 × TBE for 1 h and photograph the gel on a UV illuminator.
4. Repeat the *Sau*3AI digestion condition that produced the desired partial digestion[e] using 500 μl of genomic DNA. After stopping the reaction, layer it on to a 36 ml 10–40% (w/v) sucrose gradient (in 20 mM Tris–HCl pH 8.0, 0.8 M NaCl, 10 mM EDTA) and centrifuge it at 18 000 r.p.m. for 20 h at 4°C in an SW28 rotor (or its equivalent).
5. At the end of the run puncture the tube at the bottom and collect 70, 0.5 ml fractions in 24-well tissue culture dishes. Remove 20 μl aliquots from every sixth fraction and spot dialyse them at room temperature for 1 h on Millipore filters (type VS 0.025 μm) that are floated on 20 ml of TE. Remove the samples from the filter, add 2 μl 10 × stop solution, and analyse by FIGE as described in step 3. When the fractions containing DNA in the 60–140 kb range are identified, analyse 20 μl aliquots from every fraction in the region of the gradient that contains these DNA fragments. Decide which fractions (usually four to six) are to be processed further. Spot dialyse these fractions as described above and then concentrate them by multiple rounds of butanol extraction as follows. Add 6 vol. of NaCl-saturated butanol to the DNA sample in a 10 ml plastic tube, place the tube on its side and rock gently for 45 min. Remove the upper butanol phase. Repeat until the volume of the lower

aqueous phase is reduced to about 100 μl (should take about four or five extractions). Air dry the aqueous phase on a strip of Parafilm to remove all of the butanol, and then spot dialyse the fraction for 30 min at room temperature to remove any concentrated sucrose–NaCl. Transfer the fraction to a 1.5 ml microcentrifuge tube and repeat the butanol extractions until the volume of the aqueous phase reaches about 25 μl. Spot dialyse the sample for 1 h after evaporating any excess butanol.[f] This is the DNA that will be ligated to the p*Ad*10*sac*B vector arms.

[a] Quantifying the HMW genomic DNA can be difficult because it is so viscous and because it is non-homogeneous. It is often convenient to measure DNA concentration of an aliquot after it is digested with a restriction enzyme. Pulse field gel electrophoresis (see below) is performed on a number of different aliquots to determine the size distribution of the DNA.
[b] See also Chapter 1, *Protocol 4*; Chapter 2, *Protocol 4*; Chapter 4, *Protocol 3*.
[c] The 6 h 4°C magnesium-free, *Sau*3AI incubation is designed to produce a homogeneous distribution of the enzyme on the viscous DNA before the reaction is initiated with magnesium. This procedure produces a more uniform digestion of that DNA than is observed by initiating the reaction with enzyme.
[d] It should be noted that viscous DNA samples are difficult to load into submerged gels. Accordingly, one should load samples into the wells of a dry gel and then cover those wells with molten 1% agarose (60°C) before submerging the gel in the running buffer of the electrophoresis chamber.
[e] A desirable partial digestion has about 20% of the DNA in the 70–120 kb size range with not more than 10% of the DNA $<$ 50 kb in size.
[f] It is critical to remove all of the butanol from a sample before placing it on the dialysis filter, otherwise the sample will slide off the filter into the dialysis buffer. It is also important to analyse the final DNA preparation by FIGE to determine whether the DNA has in fact been concentrated and to make sure it has not been broken during the processing steps.

4. Preparation of P1 packaging extracts

Packaging of vector–insert DNA is a two-step process. In the first step an extract is used that contains P1 *pac* recognition and cleavage proteins (pacases). This extract lacks phage heads, tails, and accessory proteins, and is produced by thermally inducing an *E. coli* cell line with a temperature-sensitive P1 prophage that also contains a gene 10 amber mutation. Gene 10 is needed for the production of all late P1 proteins but is not necessary for the production of pacase proteins (19, 20). Once the *pac* site on the vector has been recognized and cleaved that DNA must be brought into an empty phage prohead, condensed, and a tail must be added to complete the virus. The stage II extract carries out this process. It is generated by thermally inducing a cell line with a P1 prophage that contains an amber mutation (*am*131) in one of its pacase genes. This extract produces no functional pacase protein but makes phage heads and tails. The strains used to make these extracts contain other important mutations that inactivate the P1 restriction–modification system (r^-m^-), the *E. coli* exonuclease V function (*recD*1015), the *E. coli* restriction system (*hsdR*), and the *E. coli* MeCpG restriction system (*mcrA*, *mcrB*). In this section we describe how the stage I and stage II P1 packaging extracts are prepared (*Protocol 3*).

Protocol 3. Preparation of P1 packaging extracts

Equipment and reagents

- NS3208 [=MC1061 *sup°* *recD*1014 *hsdR*$^-$ *hsdM*$^+$ *mcrA*$^-$ *mcrB*$^-$ (P1 r$^-$m$^-$ cm-2 *c*1.100 *am*10.1)][a]
- LB (see *Protocol 1*) containing 25 μg/ml chloramphenicol
- LB agar plates containing 25 μg/ml chloramphenicol
- 350 ml plastic bottles for the Sorvall GSA rotor (or equivalent)
- Pacase buffer: 20 mM Tris–HCl pH 8.0, 1 mM EDTA, 50 mM NaCl, 1 mM PMSF
- Branson sonifier
- NS3210 [=MC1061 sup° *recD*1014 *hsdR*$^-$ *hsdM*$^+$ *mcrA*$^-$ *mcrB*$^-$ (P1 r$^-$m$^-$ cm *c*1.100 *am*131)]
- 50% sucrose
- 50 mM Tris–HCl pH 8.0, 10% sucrose
- Liquid nitrogen
- Lysozyme: a freshly prepared solution at 10 mg/ml

A. *The pacase or stage I extract*

1. Streak an aliquot of a culture of strain NS3208 on to an LB agar plate containing 25 μg/ml chloramphenicol. Incubate the plate overnight at 30–32°C and then pick several colonies for growth into 10 ml cultures of LB containing 25 μg/ml chloramphenicol at 32°C. When the cultures reach late exponential phase (OD_{630} = 1.0) spread 20 μl of a 1:100 dilution on to each of two LB agar–chloramphenicol plates. Incubate one at 32°C, and the other at 42°C, overnight. Put the remainder of the cultures at 4°C. The next day score the colonies on the plates and choose, for further processing, the culture that produces the highest 32°/42° ratio—it should be > 100.
2. Inoculate 1 litre of LB containing 25 μg/ml chloramphenicol with 10 ml of the chosen culture and shake it at 30°C until it reaches an OD_{630} of 0.5 (about 4–5 h). Pellet the cells by centrifugation in 350 ml plastic bottles using a Sorvall GSA rotor (or its equivalent) at 7000 r.p.m. for 10 min at 4°C. Resuspend the cell pellet in 5 ml LB, and dilute the cell suspension into 1 litre of LB containing chloramphenicol that has been pre-warmed to 42°C. Shake the culture at this temperature for 15 min, lower the temperature to 38°C, and continue shaking for 165 min.
3. Chill the culture rapidly to 4°C by shaking it for several minutes in a water–ice slurry. Pellet the cells by centrifugation as described in step 2. Resuspend the pellet in 2 ml of cold (4°C) pacase buffer.
4. Sonicate the resuspended cell suspension with the fine tip of a Branson sonifier at setting 5 for four 15 sec intervals at 4°C.[b]
5. Centrifuge the sonicated extracts for 30 min at 17 000 r.p.m. in a Sorvall SS34 rotor (or its equivalent) and then distribute 20 μl aliquots of the supernatant to separate 0.5 μl microcentrifuge tubes and store at

−70°C. The extract is stable for as many as five rounds of freeze-and-thaw if kept at 4°C when thawed.

B. *The head–tail or stage II extract*

1. Streak out an aliquot of strain NS3210 and process as described for strain NS3208 above. Harvest the 1 litre of cells grown at 32°C when it reaches OD_{630} = 0.3. Resuspend the pelleted cells in 5 ml LB. Dilute the resuspended pellet into 500 ml LB–chloramphenicol pre-warmed to 42°C, and culture at 42°C with rigorous shaking (300 r.p.m. in an air shaker) for 20 min. Continue to shake at 38°C for an additional 25 min.[c]
2. Remove the culture from the shaker, add 500 ml ice-cold 50% sucrose, and swirl it for 1–2 min in an ice water-bath. Pellet the cells in pre-cooled 350 ml plastic bottles in a GSA rotor (or equivalent) at 7000 r.p.m. for 5 min at 4°C. Pour off all of the supernatant and dry the inside of the bottle with a tissue.
3. Resuspend the pellet in 2 ml ice-cold 50 mM Tris–HCl pH 8.0, 10% sucrose by mixing with the tip of a pipette. If the suspension is not viscous, aliquot 50 μl aliquots into microcentrifuge tubes containing 4 μl of a 10 mg/ml solution of lysozyme at 4°C. If the resuspended cell pellet is viscous (indicating some cell lysis) sonicate the suspension for two or three 10 sec intervals at 4°C as described in *Protocol 3A*, step 4, and aliquot as described above. Quick freeze all of the tubes containing the 50 μl aliquots in liquid nitrogen and store at −70°C until ready to use.

[a] *c*1.100 is a temperature-sensitive P1 repressor mutation, cm−2 is a tn9 transposon in P1 DNA that encodes chloramphenicol-resistance, and sup° indicates the absence of a nonsense suppressor.

[b] It is often difficult to duplicate sonication conditions exactly. Thus, it is best to sonicate the cells for two, four, and six 15 sec intervals, and then use the extract that produces the best results in subsequent packaging reactions.

[c] It should be noted that as NS3210 does not contain a lysis defective P1 prophage (the 10.1 amber mutation in NS3208 inhibits phage lysozyme production). Thus, care must be taken to harvest the cells rapidly no later than 45 min after induction to avoid cell lysis during subsequent processing steps. The key to the procedure is to cool the cells quickly, keep all containers cold, and to add sucrose to the cell suspension before centrifugation to maintain cell integrity.

5. Cloning of high molecular weight DNA inserts

5.1 Generation and characterization of P1 clones

The following four protocols describe how starting with vector DNA, vector arms are generated by restriction digestion and phosphatase treatment (*Protocol 4*). The arms are then ligated to size selected, *Sau*3AI digested genomic DNA (*Protocol 5*). The product is incubated consecutively with two different

P1 packaging extracts to produce virus particles containing the vector with its cloned insert (*Protocol 6*). The phage are used to infect *recA*$^{-}$, restriction deficient, Cre^{+} bacteria and P1 clones are selected on agar plates with kanamycin and sucrose (*Protocol 7*).

Protocol 4. Preparation of vector arms

Reagents

- p*Ad*10*sac*BII vector DNA (see section 3.1)
- T4 DNA ligase (NEB#202S)
- 10 × T4 DNA ligase buffer (New England Biolabs buffer with ATP)
- *Bam*HI and *Sca*I (NEB#136S and NEB#122S)
- Bacterial alkaline phosphatase (BAP; BRL #80115A)
- 50 mM Tris–HCl pH 8.0, 0.1 mM EDTA
- TE (see *Protocol 1*)
- 10 X stop solution (see *Protocol 2*)
- Calf intestinal alkaline phosphatase (New England Nuclear CIP; NEE-120)
- 3 M sodium acetate pH 5.2

Method

1. Digest 5 μg of p*Ad*10*sac*BII DNA with 10 U of *Sca*I at 37°C for 2 h in a 50 μl reaction as specified by the vendor. To ensure complete digestion repeat the digestion with an additional 10 U of *Sca*I.
2. Increase the volume of the reaction to 100 μl by adding 50 μl 10 mM Tris–HCl pH 8.0, 1 mM EDTA (TE). Extract the DNA with 1 vol. of phenol and then 1 vol. of chloroform/isoamyl alcohol (24:1). Remove the aqueous phase and add 10 μl of 3 M sodium acetate and 2 vol. of 100% ethanol. Leave at −20°C for 2 h to precipitate the DNA. Pellet by microcentrifugation. Wash the pellet with 75% ethanol and allow to air dry (or dry in Speed-vac). Resuspend the pellet in 50 μl of 50 mM Tris–HCl pH 8.0, 0.1 mM EDTA. Remove a 1 μl aliquot and dispense into 20 μl TE. Add 2 μl 10 × stop solution, heat the sample to 70°C for 10 min, and store at 4°C.
3. Treat the *Sca*I blunt-ended DNA with 500 U bacterial alkaline phosphatase (BAP) at 65°C for 2 h. Extract the DNA with phenol/chloroform and precipitate with ethanol as in step 2. Resuspend the pellet in 50 μl 1 × *Bam*HI buffer (a 1:10 dilution of 10 × NEB *Bam*HI buffer). Remove and treat a 1 μl aliquot as described in step 2.
4. Digest the DNA with 10 U of *Bam*HI at 37°C for 60 min. Extract the DNA as described in step 2 and resuspend the pellet in 50 μl of 50 mM Tris–HCl pH 8.0, 0.1 mM EDTA. Remove and treat a 1 μl aliquot as described in step 2.[a]
5. Dilute the calf intestinal alkaline phosphatase 1:2000 in water and add 1 μl of the dilution to the DNA. Incubate the reaction at 37°C for 1 h. Extract and precipitate the DNA as in step 2, but use two separate phenol extractions, and resuspend the vector arm DNA in 25 μl TE (200 ng/μl). Remove a 1 μl aliquot and treat as described in step 2.[b]

6. The four 1 μl aliquots generated in steps 2–5 are analysed by field inversion gel eletrophoresis (FIGE) as described in *Protocol 2*. The DNA from steps 2–3, should produce a single DNA fragment that migrates at 30.5 kb in size. The DNAs from steps 4–5 should produce two DNA fragments that migrate at 26.2 kb and 4.3 kb. If the *Bam*HI or *Sca*I digestion is incomplete a third band migrating at 30.5 kb will also be seen. Digestion should be at least 90% complete before proceeding with *Protocol 5*.

[a] The DNA should not be digested more extensively with *Bam*HI as this is associated with secondary nuclease or star activity. Thus, we have noted an unacceptable increase in false positives (sucrose-resistant colonies) when the *Bam*HI cloning site in the *sac*BII vector is digested for more than 1 h.
[b] Treating the DNA with more CIP, or for a longer time with CIP, reduces the efficiency of cloning. The CIP treatment described is designed to reduce ligation efficiency of the *Bam*HI ends about tenfold.

Protocol 5. Ligation of vector arms to *Sau*3AI digested and size fractionated genomic DNA

Reagents

- Vector arm DNA (see *Protocol 4*)
- Size selected, *Sau*3AI digested genomic DNA (see *Protocol 2*)
- 50 mM ATP solution pH 7.5
- T4 DNA ligase (NEB#202S)
- 10 × T4 DNA ligase buffer (New England Biolabs buffer with ATP)

Method

1. Add 1 μl vector arm DNA (200 ng) to 7 μl (500 ng) of size selected, *Sau*3AI digested genomic DNA[a] (see *Protocol 2*).
2. Incubate the mixture for 5 min at 65°C in a water-bath and then allow the sample to slowly cool to 30°C by turning the water-bath off.[b]
3. Add 1 μl 10 × ligase buffer (NEB) and 1 μl T4 DNA ligase. Gently mix the sample with a Gilson pipette tip and incubate it at 16–18°C overnight.
4. Inactivate the reaction by heating it to 70°C for 5 min. It can then be stored at 4°C for several weeks without loss of activity.

[a] It should be noted that high molecular weight genomic DNA or its ligation products should always be handled gently. *Do not vortex*. Moreover, when pipetting the DNA, it is prudent to cut the end of the Gilson tip to increase its bore size. This minimizes shear forces.
[b] Pre-heating the vector–insert mixture prior to ligation denatures any pre-annealed single-stranded *Bam*HI or *Sau*3AI ends, and can increase cloning efficiency as much as fivefold.

Protocol 6. Packaging the ligated DNA

Reagents

- Ligated DNA (from *Protocol 5*)
- P1 packaging extracts (see sections 4.1, 4.2)
- 10 × Tris–NaCl–$MgCl_2$: 100 mM Tris–HCl pH 8.0, 250 mM NaCl, 100 mM $MgCl_2$
- 1 × stage II packaging buffer: 6 mM Tris–HCl pH 8.0, 60 mM $MgCl_2$, 60 mM spermidine, 60 mM putrescine, 30 mM β-mercaptoethanol
- 30 mM DTT
- 10 × dNTPs: 1 mM each dATP, dGTP, dCTP, dTTP
- TMGD: 10 mM Tris–HCl pH 8.0, 10 mM $MgCl_2$, 0.1% gelatin, 10 μg/ml pancreatic DNase (the DNase should be added to TMG from a 2 mg/ml solution just prior to use)

Method

1. Heat the ligated DNA reaction to 70°C for 10 min and place on ice. Add 3 μl of this reaction to 1.5 μl 10 × Tris–NaCl–$MgCl_2$, 1.5 μl 10 × dNTPs, 1 μl 30 mM DTT, 0.5 μl 50 mM ATP, 6.5 μl water, and 1 μl stage I P1 packaging extract. Incubate the reaction for 15 min at 30°C.[a]
2. Add 3 μl of stage II buffer and 1 μl of 50 mM ATP to the stage I reaction and pipette the entire solution into a tube containing the stage II head–tail extract (about 40–50 μl). Mix the resulting extract with the tip of a Gilson pipette (it should be viscous) and then incubate it at 30°C for 20–30 min.
3. Terminate the packaging reaction by adding 120 μl of TMGD buffer and continue the incubation for 15 min at 37°C. The reaction should become significantly less viscous.
4. Centrifuge the reaction for 1 min in a microcentrifuge to pellet debris. Remove the supernatant (about 180 μl) to a separate tube, and store at 4°C until used.[b]

[a] It is convenient to pre-mix the 10 × Tris–NaCl–$MgCl_2$, 10 × dNTPs, DTT, and ATP, and add them to the reaction together. It is best to add the stage I extract last, and the reaction should not be incubated longer than 15 min as this tends to decrease the efficiency of the process due to trace nucleases in the extract.

[b] While the phage lysate can be stored at 4°C for several weeks, it is best to use it as soon as possible to maximize recovery of P1 clones. If it has to be stored more than a few days it is best to add 15 μl of chloroform to the reaction after step 4, vortex, and proceed as described.

Protocol 7. Generation of P1 clones

Equipment and reagents

- Luria broth (LB) (15)
- LB agar (15)
- LB with 5 mM $CaCl_2$
- LB agar plates containing either 25 μg/ml kanamycin or 25 μg/ml kanamycin plus 5% sucrose
- NS3529 = *E. coli* K12 *recA mcrA Δ* (*hsd R hsd M mcrB mcrC mrr*) (λimm434*nin*5X1-*cre*) (λimmλLP1)[a]
- RT6000 centrifuge (or its equivalent)
- Packaging reaction (from *Protocol 6*)

Method

1. To prepare bacteria for infection with packaged P1 phage grow a culture of *E. coli* strain NS3529 in Luria broth (LB) at 37 °C to mid-log phase (OD_{590} = 0.6) starting from a colony on an LB agar plate. Pellet the cells by centrifugation using a Sorvall H1000B rotor in an RT6000 centrifuge (or its equivalent) for 10 min at 3500 r.p.m. at 4 °C. Resuspend the pellet in one-tenth culture volume LB containing 5 mM $CaCl_2$.
2. Mix up to 30 µl of the packaging reaction with 200 µl of the NS3529 cell suspension and incubate the mixture at 37 °C for 5 min. Add 1 ml LB and incubate the culture at 37 °C with gentle shaking for an additional 30–45 min.[b]
3. Pellet the P1-infected culture in a microcentrifuge for 2 min at room temperature and then resuspend the pellet in 200 µl LB. Spread aliquots on LB agar plates containing either 25 µg/ml kanamycin or 25 µg/ml kanamycin plus 5% sucrose. Incubate the plate at 37 °C overnight and score the colonies.

[a] The *imm*434 prophage produces Cre, constitutively (21), and the *imm*λ prophage produces the *lacI*^{q} repressor, constitutively.
[b] It is important not to extend the growth period for more than 45 min to minimize sibling production. Also, if the phage supernatant has been exposed to chloroform it is best to let the 30 µl aliquot air dry for 10–15 min to allow the chloroform to evaporate before adding the cells. Finally, it is best not to use more than 30 µl of packaged phage/0.2 ml of cells as the yield of transformed colonies is no longer linear with increasing phage volume beyond this point.

5.2 Typical cloning results

Typical cloning results are shown in *Table 1*. With just vector DNA or vector DNA digested with *Bam*HI and ligated to produce concatemers, the packaging reaction generates $4 \times 10^5 - 2 \times 10^6$ kanamycin-resistant colonies per microgram of vector (lines 1, 3). However, only a few per cent of these colonies grow on agar plates with sucrose. The unligated vector arms are packaged less than 30 × as efficiently as is uncut vector and ligation of these arms increase packaging efficiency about six- to tenfold (lines 4, 5). Again, few of these packaged phage yield sucrose-resistant colonies. If the arms are ligated along with insert DNA then the phage produced generate an additional twofold more kan^r cells but now 70% of these transformants are also resistant to sucrose and have genomic inserts (line 6). Note, failure to detect at least a five- to tenfold increase in kan^r transformants when vector arms are ligated indicate that they have probably been treated too extensively with CIP. They will yield less than an optimal number of P1 clones when genomic DNA is added to the reaction.

5.3 Isolation and analysis of P1 plasmid DNA

This section describes the alkaline lysis procedure used to isolate P1 plasmid DNA on a small scale and the restriction analysis of that DNA.

Protocol 8. Isolation and analysis of P1 plasmid DNA

Equipment and reagents

- 0.1 M IPTG solution
- TGE: 25 mM Tris–HCl pH 8.0, 50 mM glucose, 10 mM EDTA
- 0.2 N NaOH, 1% SDS
- Potassium acetate: 3 M potassium, 5 M acetate, pH 4.8 (18)
- 2 mg/ml RNase I (pre-incubated 5 min at 95°C)
- LB with kanamycin (from *Protocol 1*)
- LB with kanamycin and sucrose (from *Protocol 7*)
- Type VS dialysis filters (Millipore, 0.025 μm)
- *Bgl*III and *Not*I (NEB#144S and NEB#1895)
- Lysozyme (10 mg/ml), freshly prepared

A. *Preparation of P1 plasmid DNA*

1. Inoculate 10 ml LB containing 25 μg/ml kanamycin with a single colony from an LB agar kan–suc plate. Culture with vigorous aeration at 38°C until the cells reach early exponential phase (OD_{590} = 0.2).
2. Add 100 μl of 0.1 M IPTG to the culture and culture for an additional 3–4 h at 38°C.
3. Pellet the cells by centrifugation, resuspend in 150 μl TGE, and add 15 μl of a 10 mg/ml lysozyme solution. Transfer to a 1.5 ml microcentrifuge tube and keep at room temperature for 10 min.
4. Add 300 μl of a fresh 0.2 M NaOH, 1% SDS solution. Invert the tube several times and keep at 4°C for 10 min.
5. Add 225 μl potassium acetate, invert tube again several times, and keep on ice for an additional 5 min.
6. Centrifuge the lysate for 5 min at 12 000 r.p.m. in an SS34 rotor (RC-5C Sorvall centrifuge) at 4°C and transfer the supernatant to a tube containing 0.5 ml phenol. Invert the tube several times, and centrifuge for 2 min in a microcentrifuge. Remove the lower phenol phase. Add 0.5 ml chloroform/isoamyl alcohol (24:1) to the aqueous phase and repeat the above extraction. Remove upper (aqueous) phase and precipitate nucleic acids by adding 2 vol. (1 ml) ethanol, and leaving for at least 1 h at −20°C.
7. Pellet the precipitate by centrifuging. Wash the pellet with 75% ethanol. Dry the pellet in a Speed-vac and resuspend in 40 μl TE containing 0.5 μg RNase I.
8. Spot dialyse (see *Protocol 2*) the DNA solution on a type VS dialysis filter against 100 ml TE for 30 min and store at 4°C. This procedure yields about 1–2 μg of plasmid DNA or enough DNA for about four digests. It can be efficiently scaled up to 250 ml using the following reagent volumes: 2.5 ml TGE, 4.5 ml NaOH–SDS solution, 3.3 ml potassium acetate. The DNA pellet is resuspended in 400 μl TE with 4μg RNase.

B. *Restriction analysis of P1 plasmid DNA*

1. Remove 10 μl from the isolated DNA sample and digest it with either 5 U of *Bgl*III or 5 U of *Not*I in a 20 μl reaction at 37 °C for 2 h using conditions as specified by the vendor (New England Biolabs).[a]
2. Add 2 μl 10 × stop solution and heat the reactions for 5 min at 70 °C.
3. Analyse the *Bgl*II digests by standard agarose gel electrophoresis and the *Not*I digests by FIGE (see *Protocol 2*).

[a] Digestion of P1 vector–insert DNA with *Bgl*II produces a 6.0 kb fragment from the vector. The other fragments in the digest are from the insert sequences or are junction fragments. Since *Not*I cuts the vector once it should produce a single fragment if there are no *Not*I sites in the insert. *Not*I and *Sal*I double digest will cleave the insert sequences away from the vector and will produce a unique 16–17 kb DNA vector fragment.

6. Trouble-shooting

6.1 Vector DNA

The two problems that are most frequently encountered are the presence of vector DNA with a defective *sac*B gene, and inappropriately prepared vector arms. The first problem was discussed in section 3.1. The second problem is most usually due to over-treatment of the vector arms with CIP in the last step of their preparation. This can be evaluated by comparing the packaging efficiency of ligated and unligated arms (*Table 1*). Ligation should increase the packaging efficiency about six to tenfold (lines 4, 5). Another potential problem is incomplete digestion of vector DNA with either *Sca*I or *Bam*HI. Analysis of the products of each step in the preparation of the arms and repeating those steps, if necessary, is desirable. If *Sca*I digestions are not carried out under appropriate conditions fragments other than those expected are produced. We assume these are due to a 'star' activity.

6.2 *Sau*3AI digested, size fractionated insert

If vector arms satisfy the criteria shown in *Table 1*, lines 4, 5, the failure to clone insert fragments efficiently (*Table 1*, line 6) must indicate either that the *Sau*3AI ends of the insert have been damaged (likely broken) or, more likely, that the concentration of insert is too low. The latter problem is overcome by starting the butanol concentration process with more size fractionated insert DNA. To accomplish this one should carry out more sucrose gradients rather than adding more DNA to each gradient. Addition of more than 50 μg of digested DNA to the gradient tends to overload it, reducing the efficiency of size fractionation.

6.3 Packaging extracts

The extracts are tested most easily using uncut vector DNA as a substrate. Packaging efficiency with this DNA substrate is about 10% that observed with concatemeric vector DNA. Failure to achieve the expected efficiency is most easily dealt with by repeating the preparation, as described, with a different colony isolate.

7. Anticipated results and time considerations

As indicated by the results shown in *Table 1* each packaging reaction should generate about 12 × 400 clones, or about 5000 clones. Since only 3 μl of the 10 μl ligation reaction is used per packaging reaction about 15 000 clones will be generated per ligation reaction (200 ng vector arms, 500 ng of insert). Thus a library giving a three times coverage of the mouse or human genomes (about 130 000, 80 kb insert clones) should be generated by 30 packaging reactions, 10 ligation reactions, and 2 μg vector arms plus 4 μg of insert fragment. The entire process of preparing vector, packaging extracts, inserting DNA, and testing each these components will take one person about two months. The preparation of packaging extracts, plating out, and pooling of clones (see below) will take another month. Thus, if all goes well a library can be prepared in about three months.

8. Organizing and screening a P1 library

A convenient method of organizing a P1 library is described in ref. 1. Briefly, a mouse library of about 130 000 clones with an average insert size of 75 kb is organized in 300 primary pools of 400–500 clones each. Cells from the primary pools are mixed in groups of ten, amplified, and plasmid DNA isolated. These constitute 30 secondary pools of 4000–5000 clones each. DNAs from five secondary pools are pooled to generate six tertiary pools of 25 000 clones each. The pools are screened in a top down approach by PCR analysis of DNA from tertiary, then secondary, and finally primary pools. Once a positive primary pool is identified, the desired clone in the pool is selected by colony hybridization using the PCR product as probe.

Alternatively, the clones in the library may be individually arrayed, as was the case in the construction of human P1 library (22). Screening of this library (22) is also by PCR using various pooling strategies.

References

1. Pierce, J., Sternberg, N., and Sauer, B. (1992). *Mamm. Genome*, **3**, 550.
2. Foote, S., Vollrath, D., Hilton, A., and Page, D. (1992). *Science*, **258**, 60.
3. Gasser, D., Sternberg, N., Pierce, J., Goldner-Sauve, A., Feng, H., Haq, A., *et al.* (1993). *Immunogenetics*, **39**, 48.
4. Sternberg, N. and Coulby, J. (1987). *J. Mol. Biol.*, **194**, 469.
5. Sternberg, N. and Coulby, J. (1990). *Proc. Natl Acad. Sci. USA*, **87**, 8070.
6. Austin, S., Hart, F., Abeles, A., and Sternberg, N. (1982). *J. Bacteriol.*, **152**, 63.

7. Sternberg, N. and Cohen, G. (1989). *J. Mol. Biol.*, **207**, 111.
8. Sternberg, N., Ruether, J., and DeRiel, K. (1990). *New Biol.*, **2**, 151.
9. Gay, P., Lecog, D., Steinmetz, M., Ferrari, E., and Hoch, J. A. (1983). *J. Bacteriol.*, **153**, 1424.
10. Tang, L. B., Lenstra, R., Borchert, T. V., and Nagarajan, V. (1990). *Gene*, **96**, 89.
11. Pierce, J., Sauer, B., and Sternberg, N. (1992). *Proc. Natl Acad. Sci. USA*, **89**, 2056.
12. Eliason, J. L. and Sternberg, N. (1987). *J. Mol. Biol.*, **198**, 281.
13. Raleigh, E. A. and Wilson, G. (1986). *Proc. Natl Acad. Sci. USA*, **83**, 9070.
14. Raleigh, E. A., Murray, N. E., Revel, H., Brumenthal, R. M., Westaway, D., Reith, A. D., *et al.* (1988). *Nucleic Acids Res.*, **16**, 1563.
15. Maurer, R. and Sternberg, N. (1991). In *Methods in enzymology* (ed. J. Miller), Vol. 204, pp. 18–43.
16. Dagert, M. and Ehrlich, S. D. (1979). *Gene*, **6**, 23.
17. Sternberg, N. (1992). In *Methods in enzymology* (ed. R. Wu), Vol. 216, pp. 549–
18. Sambrook, J., Fritsch, E. F., and Maniatis, T. (ed.) (1989). *Molecular cloning, a laboratory manual* (2nd edn). Cold Spring Harbor Press, Cold Spring Harbor, NY.
19. Walker, J. T. and Walker, D. H. (1983). *J. Virol.*, **45**, 1118.
20. Sternberg, N. and Coulby, J. (1987). *J. Mol. Biol.*, **194**, 453.
21. Sauer, B. and Henderson, N. (1988). *Gene*, **70**, 331.
22. Shepherd, N., Pfrogner, B. D., Coulby, J. N., Ackerman, S. L., Vaidyanathan, G., Sauer, R., *et al.* (1994). *Proc. Natl Acad. Sci. USA*, **91**, 2629.

4

Cloning into yeast artificial chromosomes

RAKESH ANAND

1. Introduction

The analysis of large and complex genomes requires both mapping and cloning of DNA. Until a few years ago the largest fragment of DNA that could be cloned was ~ 40 kb in length using a cosmid vector (see Chapters 1 and 2). The development of yeast artificial chromosome (YAC) cloning vectors has greatly extended this cloning range. Following the description of the YAC vector system (1) the size of DNA fragment that could be maintained as a YAC has increased by almost one order of magnitude. This has greatly facilitated the study of complex genomes and has been instrumental in the efforts to construct the first generation physical map of the entire human genome (2).

A well constructed primary YAC library with clones ordered into microtitre plates provides a valuable long-term resource which can be simultaneously accessed by several researchers. Consequently, it is important to plan carefully before embarking on this exercise that requires substantial time and resource. This chapter will mainly concentrate on the construction of a YAC library. For completeness, determination of the average YAC size within the library and a PCR-based library screening method have also been included.

2. YAC vectors

Methods described in this chapter were developed using pYAC4 (*Figure 1*) as the cloning vector (1) and hence the protocols below will refer only to this vector. However, several other vector and host strains have now been developed offering significant advantages over pYAC4 (see below) and should be considered as serious alternatives. This may require some minor modification to the protocols.

The vector pYAC4 (1) when restricted with *Bam*HI and *Eco*RI produces three fragments (*Figure 1*). The two larger fragments form the YAC arms

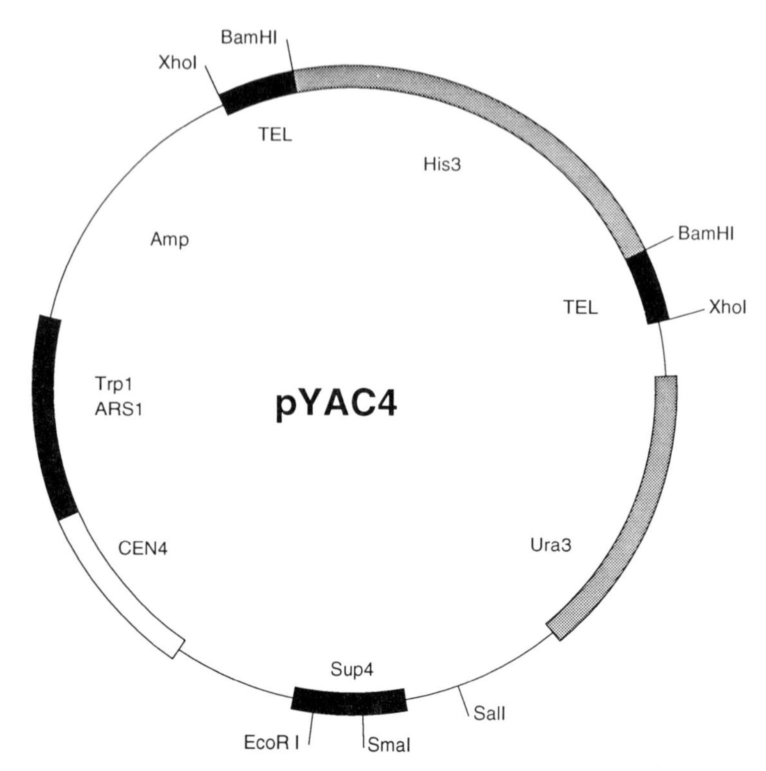

Figure 1. Schematic drawing (not to scale) to illustrate one of the YAC cloning vectors pYAC4. The cloning site *Eco*RI is within the *Sup4* gene. *Trp1* and *Ura3* are the left and right YAC arm yeast selectable markers respectively. *TEL* are the two telomere seeding sequences and *CEN* provides the centromere function in *Saccharomyces cerevisiae*.

with the *Bam*HI ends of the *TEL* sequences providing seeding sequences for the formation of functional telomeres in *Saccharomyces cerevisiae*. The third fragment, which contains the *His3* gene, is a 1.8 kb 'stuffer' fragment which does not participate in the cloning. The 6 kb (left) arm contains an *ARS1* sequence enabling autonomous replication, *CEN4* sequence providing all the *cis*-acting elements for mitotic and meiotic centromere function, and a marker gene *Trp1* which can be used for positive selection. The 3.4 kb (right) arm has the *Ura3* gene which can also be used for positive selection. Finally, the *Eco*RI cloning site is within the *Sup4* gene and when a DNA fragment is cloned within it, the ochre suppression is eliminated. Therefore, in an *Ade2*-ochre host, adenine metabolism is interrupted resulting in the accumulation of a red pre-metabolite (phosphoribosylaminoimidazole) providing recombinant clones with their diagnostic red colour.

Some of the alternatives to pYAC4 include a two vector system (3) incorporating the advantages of multiple cloning sites, T7 bacteriophage promoters for the generation of riboprobes, and the potential for YAC end rescue as plasmids. Furthermore, YACs constructed using these vectors can be fragmented or modified to incorporate a mammalian selectable marker using complementary vectors developed by the same group (4, 5). Another YAC vector pCGS966 allows the amplification of the artificial chromosome in yeast (6).

2.1 Preparation of YAC vectors

This is an extremely important step since the integrity of the restricted and dephosphorylated ends of the vector arms have a very significant effect on the cloning efficiency and the level of background colonies. Although the method described below is for pYAC4, it can easily be adapted for other YAC vectors. pYAC4 has a single *Eco*RI cloning site and can be used with the host *Saccharomyces cerevisiae* AB1380 (*MATa*ψ^+, *ura3, trpl, ade2–1, canl–100, lys2–1, his5*).

Protocol 1. Vector preparation

Equipment and reagents

- Restriction enzymes *Eco*RI and *Bam*HI and the manufacturer's recommended buffers (core buffers 'B' or 'C' from Promega can be used for both enzymes)
- Calf intestinal alkaline phosphatase and the manufacturer's recommended buffer
- T4 DNA ligase and the manufacturer's recommended buffer
- 7.5 M ammonium acetate (sterilize by autoclaving)
- 0.5 M EDTA pH 8.0 (sterilize by autoclaving)
- TAE (50 X) concentrated stock solution: 242 g Tris base, 57.1 ml glacial acetic acid, 100 ml 0.5 M EDTA pH 8.0 per litre final volume
- TE: 10 mM Tris–HCl pH 8.0, 1 mM EDTA (sterilize by autoclaving)
- Agarose (electrophoresis grade)
- Chloroform/isoamyl alcohol (24:1 v/v)
- Phenol/chloroform/isoamyl alcohol (25:24:1 by vol.) equilibrated with TE (Fisons or similar supplier of Molecular Biology reagents)
- Agarose gel electrophoresis apparatus and reagents
- L broth: 10 g bacto tryptone, 5 g bacto yeast extract, 10 g NaCl per litre; adjust to pH 7–7.2 and sterilize by autoclaving
- L broth + amp (25–50 μg/ml ampicillin from a stock of 50 mg/ml)

Method

1. Prepare more than 500 μg vector DNA using a standard maxi-prep method (7). pYAC4 is a pBR322-based plasmid which can be grown in *E. coli* host strain HB101 (*recA*). Due to the large size of this plasmid the yield is low (100–200 μg/litre L broth + amp).
2. Restrict 500 μg vector DNA with *Bam*HI in the appropriate buffer and test the digest for completion, i.e. the presence of two fragments (9.4 kb and 1.8 kb) by agarose gel electrophoresis.
3. Restrict with *Eco*RI. If Promega enzymes and core buffers are used, this step can be carried out by simply adding *Eco*RI to the *Bam*HI restricted DNA from step 2. Alternatively, add 0.5 vol. of 7.5 M ammonium

Protocol 1. *Continued*

acetate and precipitate the DNA with 2 vol. of ethanol. Allow the precipitate to form at −20°C. Pellet by centrifugation and resuspend the DNA in TE. Digest with *Eco*RI in the appropriate buffer. Test the digest for completion, i.e. the appearance of three fragments (6.0 kb, 3.4 kb, 1.8 kb) upon agarose gel electrophoresis.

4. Recover the DNA by ethanol precipitation as in step 3 and resuspend in TE.
5. Dephosphorylate the DNA in the appropriate buffer using the minimum amount of pre-titrated calf intestinal alkaline phosphatase.
6. Add 1/20 vol. 0.5 M EDTA to stop the reaction and incubate at 65°C for 1 h to inactivate the enzyme.
7. Extract the DNA twice with phenol/chloroform/isoamyl alcohol followed by one extraction with chloroform/isoamyl alcohol. Add 0.5 vol. of 7.5 M ammonium acetate and 2 vol. of ethanol to the aqueous phase. Leave at −20°C for the precipitate to form.
8. Pellet the precipitation by centrifugation. Wash the DNA pellet twice with 1 ml of 70% ethanol. Re-pellet the DNA, decant off the 70% ethanol, and allow to air dry. Resuspend in 0.5 ml of TE.
9. Use approximately 1 μg of vector DNA to test the efficiency of dephosphorylation by the inability of the vector arms to self-ligate (7). If necessary, repeat the dephosphorylation (steps 5–8).
10. Test the integrity of the *Eco*RI ends by their ability to ligate to phosphorylated compatible ends. This can be done by ligating ~ 1 μg vector arms to ~ 5 μg *Eco*RI restricted genomic DNA. For unambiguous results extract the ligated DNA with phenol/chloroform/isoamyl alcohol, concentrate it by ethanol precipitation, and resuspend in TE prior to analysis by agarose gel electrophoresis.
11. Make up the restricted dephosphorylated vector stock to 1 μg/μl in TE and store in aliquots of approximately 100 μg at −20°C.

3. Insert DNA

Several YAC libraries have been constructed using total genomic DNA from a variety of sources ranging from *C. elegans* (8) to human (9–14) and maize (15). Whatever the source, it is essential to use high molecular weight DNA as the starting material. This can be achieved using standard pulsed field gel electrophoresis (PFGE) protocols for preparing DNA using cells embedded in low gelling temperature agarose. The method described here is for human DNA and is applicable to rodent as well as human–rodent hybrid cell lines. Although the principle is the same for all cell types, there are differences in the methods for obtaining adequate number of intact cells from different sources that are outside the scope of this chapter.

It is important to note that large DNA is very susceptible to damage by contaminants like nucleases. Clean techniques should be used throughout and the use of glass double-distilled water is recommended for all solutions including PFGE buffer.

Protocol 2. Insert DNA plugs [a]

Equipment and reagents

- Plug mould: Perspex plug moulds can be made in a laboratory workshop (*Figure 2*); other versions of plug moulds are available from suppliers of PFGE equipment and reagents
- Sterile polypropylene tubes (14 ml snap-top and 50 ml screw-top as produced by Falcon, Sterilin, or similar supplier)
- Haemocytometer and microscope
- Transfer pipettes (glass/plastic)
- pH sticks (Whatman plastic bonded pH sticks)
- Plastic insulation tape
- 0.2 μm Millipore filters and filter sterilization unit (Millipore)
- Nylon mesh tea strainer
- Glass double-distilled water (sterile)
- Dulbecco's saline (sterile) from Gibco or similar supplier
- LGT agarose: low gelling temperature agarose (Sea Plaque or Insert agarose from FMC corporation)
- NDS: 1% lauroyl sarcosine (or SDS), 10 mM Tris base pH 9.5, 0.5 M EDTA. To make 500 ml of this solution, add 93 g EDTA to approximately 300 ml distilled water. Add 0.605 g Tris base and then add NaOH pellets to increase the pH to greater than 8.0 to dissolve the EDTA. Dissolve 5 g N-lauroyl sarcosine in approximately 50 ml of distilled water and add this to the Tris, EDTA solution. Adjust the pH to 9.5 with NaOH and make up to 500 ml. Filter sterilize using a 0.2 μm Millipore filter and store at 4°C
- Pronase (Boehringer Mannheim): dissolve in NDS to 20 mg/ml, incubate at 37°C for 30 min to inactivate any trace contaminants, and use within 2–4 h (proteinase K (20 mg/ml in NDS) can be used as an alternative)

Method

1. Clean the plug mould according to the manufacturer's instructions. A Perspex plug mould (*Figure 2*) can be cleaned by boiling in 0.25 M HCl for 10 min. Rinse it several times in water that has been double distilled in glass to remove all traces of the acid.
2. Aim to start with 10^8–10^9 cells (tissue culture cells, or peripheral blood lymphocytes, or similar). Harvest the cells and wash once in Dulbecco's isotonic saline. Avoid any cell clumping. If it occurs, cells can be disrupted by vigorously flicking the tube containing the cell pellet (without saline).
3. Suspend the cells in 1 ml of Dulbecco's saline and estimate the total number of cells by counting a small aliquot using a haemocytometer. Dilute the cell suspension to 3×10^7 cells/ml (this yields plugs of about 100 μl containing approximately 10 μg DNA). If necessary, this cell concentration can be increased to 6×10^7 cells/ml to double the DNA concentration in the plugs.

Protocol 2. *Continued*

4. Make up approximately 10 ml of 1% LGT agarose solution in Dulbecco's saline. This is best done using a boiling water-bath to melt the agarose. Allow the agarose solution to cool to 37–40°C and hold it at this temperature.
5. Stick tape (plastic insulation tape or similar) to one surface of the plug mould and place it on ice with the taped surface in contact with the ice.
6. Warm the cell suspension to 37°C and mix it with an equal volume of 1% LGT agarose solution. Hold this mixture at 37°C and transfer it to the slots of the plug mould using a sterile transfer pipette.
7. Once the plugs are fully set (this takes about 10 min), remove the tape and push them out gently using a folded pH stick, a bent glass Pasteur pipette, or some similar device, into a Falcon tube containing 20 ml of NDS with 1 mg/ml pronase.
8. Incubate the plugs at 50°C overnight. Remove the solution and replace with fresh NDS and pronase. A convenient method of changing solutions is to cover the mouth of the tube containing the DNA plugs with a nylon mesh tea strainer. This way the solution can be drained out whilst retaining the plugs in the tube. Continue the incubation at 50°C for another 24 h. Rinse the plugs twice for 1–2 h in NDS and store in NDS at 4°C. These DNA plugs should be stable at 4°C for several years.

[a] See also Chapter 1, *Protocol 3*; Chapter 7, *Protocol 1*.

3.1 Restriction digest of insert DNA

The cloning site of the YAC vector determines the type of restriction digest for the insert, i.e. partial or complete, and if partial, the extent of the partial digest. Since the vector used in this chapter is pYAC4, the protocol below will concentrate on *Eco*RI partial restriction digest. However, this protocol can easily be adapted for partial restriction digests using other enzymes or complete restriction digests using enzymes like *Not*I that cut infrequently.

For partial restriction digests it is essential to first determine the amount of a particular batch of restriction enzyme required to generate DNA fragments such that 80–95% of the restricted genomic DNA is greater than the target size of the average YAC insert. Having determined the enzyme concentration required for the optimum partial digest in 30 min at 37°C, proceed to conduct a large scale partial digest of the insert DNA for PFGE fractionation. In order to improve the YAC library representation a range of partial digests should be included in the final DNA fractionation. This can be done by using the predetermined enzyme concentration and varying the incubation time (see *Protocol 3B*, step 5).

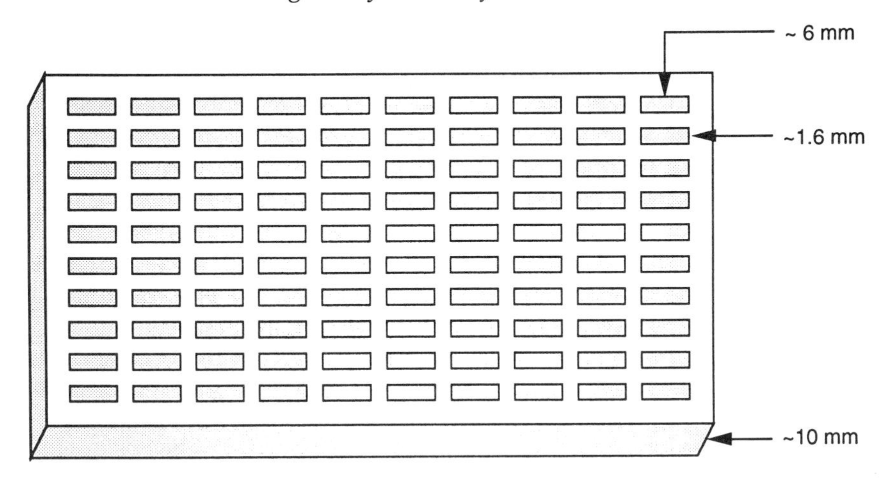

Figure 2. A 100-slot Perspex plug mould used for preparing DNA embedded in LGT agarose. The approximate dimensions of the slots are 6 mm × 1.6 mm × 10 mm resulting in ~ 100 μl agarose plugs.

Protocol 3. Restriction digest of insert DNA [a]

Equipment and reagents

- TE (see *Protocol 1*)
- *Eco*RI restriction enzyme and buffer (see *Protocol 1*)
- TAE (see *Protocol 1*)
- NDS (see *Protocol 2*)
- 50 ml and 14 ml sterile tubes (see *Protocol 2*)
- Sterile scalpel blade
- PFGE equipment and reagents

A. *Optimum enzyme concentration for a partial restriction digest*

1. Take four complete plugs. Using a clean scalpel blade, cut each one across its width into three equal portions. Wash the resulting 12 DNA plugs overnight in 50 ml TE in a sterile Falcon tube, Petri dish, or similar vessel at 4°C. Gently agitate the TE to equilibrate the plugs (this is best done on a slow roller or orbital shaker).
2. Wash the plugs for a further 2 × 1 h in 50 ml TE at 4°C.
3. Transfer individual DNA plugs into sterile microcentrifuge tubes and equilibrate with 100 μl of restriction buffer for 30 min on ice. Number the tubes from 1–12.
4. Replace the restriction buffer with 100 μl of restriction buffer containing *Eco*RI restriction enzyme (1 U/ml to 10 U/ml of *Eco*RI in increments of 1 U in tube numbers 1–10). Add 100 μl restriction buffer without any enzyme to tubes 11 and 12. Leave all tubes on ice for 30 min.
5. Incubate tubes 1–11 at 37°C for 30 min, and then add 100 μl NDS to each, and put them all on ice for at least 10 min.

Protocol 3. *Continued*

6. Wash plugs 1–12 twice in 1 ml TE and analyse by PFGE. The optimum partial digest is one where 80–95% of the restricted genomic DNA is greater than the target size of the average YAC insert. DNA plug number 11 provides a control for non-specific degradation and should not show more than a very light smear. DNA plug number 12 should not show any smear as this is a 'no incubation' control.

B. *Partial restriction digest*

1. Take ten complete DNA plugs (approximately 100 μg DNA) and wash overnight in 50 ml TE in a sterile Falcon tube, Petri dish, or similar vessel at 4°C. Gently agitate on a slow roller or orbital shaker to equilibrate the plugs.
2. Wash the plugs for a further 2 × 1 h in 50 ml TE at 4°C.
3. Transfer all ten plugs to a 14 ml sterile tube and equilibrate in 10 ml restriction enzyme buffer on ice for 30–60 min.
4. Replace the restriction buffer with 3 ml of restriction buffer containing the appropriate amount of restriction enzyme as determined by the partial digest studies above (generally this should be in the range of 1–10 U/ml). Leave on ice for 30 min.
5. Start the incubation at 37°C. Remove one plug every 5 min and place into a 50 ml Falcon tube containing 20 ml NDS on ice. Approximately 30 min after all the DNA plugs have been transferred into NDS, wash them 1 × 30 min with 50 ml of 0.5 × TAE and store in approximately 10 ml of 0.5 × TAE on ice for preparative PFGE.

[a] See also Chapter 7, *Protocol 4*; Chapter 2, *Protocol 4*; Chapter 4, *Protocol 3*.

3.2 Preparative PFGE fractionation

In the past, the manipulation of large DNA fragments in liquid, whilst keeping them intact has presented significant problems. Consequently, the earliest YAC libraries had an average insert size of 100–200 kb (9, 16, 17). Development of protocols using PFGE fractionation and careful liquid DNA handling (18) have resulted in a substantial increase in the average insert size (11–14, 19). Before aiming for a certain insert size one must decide on the major use of the YAC library. If the isolated clones are going to be used for a detailed structural and functional analysis of cloned DNA, then YAC inserts of up to ~ 500 kb provide a good substrate for accurate restriction mapping, identification of coding sequences (20), and transfer into mammalian cells (21, 22). However, if the primary purpose is genomic 'walking' to cover large distances, then inserts > 500 kb–1 Mb offer a significant advantage.

The protocol below uses a 'Waltzer' PFGE apparatus (23), available from

Hoefer Scientific Instruments under the trade name of 'Hula Gel'. However, any PFGE apparatus resulting in the migration of DNA along straight tracks can be used.

Protocol 4. Preparative PFGE of insert DNA

Equipment and reagents

- PFGE equipment and reagents
- PFGE molecular size markers (λ oligomers and *S. cerevisiae* chromosomal size markers) available from several PFGE reagent suppliers
- LGT agarose (see *Protocol 2*)
- Scalpel blade (sterile)
- TAE (see *Protocol 1*)
- 50 ml sterile tubes (see *Protocol 2*)
- pH stick (see *Protocol 2*)

Method

1. Prepare a 6–7 mm thick 1.5% agarose support gel. Once set, cut out the central portion (*Figure 3*) and pour a 1% LGT agarose gel with a single long sample loading slot (~ 1.5 mm wide and ~ 12 cm long).
2. Load the ten, *Eco*RI restricted DNA plugs laying them down along their length (i.e. 1 cm long plugs lying down in the loading slot) in the sample loading slot. Load the size markers at either end (*Figures 3* and *4*). Once all the plugs are loaded, gently push them down using a folded pH stick or similar device. Remove any excess buffer using a folded soft tissue paper and seal the DNA plugs in the loading slot using 0.5% LGT agarose solution.
3. Set the electrophoresis conditions such that only the fragments to be discarded are fractionated, i.e. if fragments below 600 kb are to be removed, PFGE conditions (pulse time, voltage, and temperature) should be such that only the fragments < 600 kb get resolved whereas the larger ones remain in the unresolved DNA compression zone. The run time for electrophoresis should be such that the DNA compression zone is within 2–3 mm of the loading slot (*Figure 4a*).
4. Following electrophoresis, excise the central portion of the gel as shown in *Figure 3* and store at 4°C.
5. Stain the rest of the gel with ethidium bromide. Visualize the region of the DNA compression zone using a UV transilluminator and make a notch in the gel to enable identification of the exact position of the DNA compression zone under normal lighting (daylight).
6. Reposition the excised LGT gel by carefully sliding it back into it's original position in the full gel. Using a sterile scalpel, cut out the region of the compression zone (by aligning it to the notch) along with the sample loading slot[a] (*Figure 4a*).
7. Cut the recovered agarose gel slice (approximately 6 mm × 9 cm) into six or seven pieces and place these in a 50 ml Falcon tube on ice.

Protocol 4. *Continued*

8. Seal the rest of the gel using 1–1.5% agarose, stain it for more than 1 h with ethidium bromide, and photograph using a UV transilluminator to confirm that the correct region of the gel has been excised. This photograph will also provide a record of the range of *Eco*RI partial digests used in the fractionation.

[a] The reason for recovering the loading well is that it contains very high molecular weight DNA which acts as a carrier. This increases the viscosity of the DNA solution in all subsequent steps, protecting the restricted fragments from damage by shear forces.

3.3 DNA purification and ligation to vector arms

During this step large DNA fragments with *Eco*RI ends are recovered in solution and ligated to the vector arms. Since large DNA fragments in liquid are liable to damage by physical shear, extreme care and gentle handling is required once the agarose gel matrix is removed and DNA is processed in solution (i.e. in step 3 and subsequent steps of *Protocol 5*).

Protocol 5. Extracting the PFGE fractionated DNA

Equipment and reagents

- Vacuum dialysis apparatus and UH 100/75 ultra thimbles available from Schleicher & Schüell (Anderman & Co. in the UK)
- Cold water tap aspirator vacuum or similar laboratory vacuum facility with a trap to prevent reverse suction
- Agarase (Calbiochem)
- 50 ml sterile tubes (see *Protocol 2*)
- TE (see *Protocol 1*)
- TE, 10 mM EDTA (from a stock of 0.5 M EDTA, see *Protocol 1*)
- TAE (see *Protocol 1*)
- High salt TAE: 1 × TAE, 100 mM NaCl, 10 mM EDTA pH 8.0 (sterilize by autoclaving)

Method

1. Equilibrate the gel slices from *Protocol 4*, step 7, with approximately 40 ml of high salt TAE on ice for 1 h.
2. Remove the buffer and replace with high salt TAE equal to the volume (weight) of the gel slice (e.g. a gel slice of 6 mm × 6 mm × 9 cm will require about 3.2 ml buffer). This high salt concentration protects DNA and prevents denaturation of DNA during the next step.
3. Heat to 68°C for 10–15 min until the agarose melts completely. Cool to about 40°C and add agarase to a final concentration of 50 U/ml. Mix gently by slowly rolling the tube on the bench and incubate at 37°C for 2–4 h with occasional mixing.
4. Add another 25 U/ml of agarase and continue the incubation at 37°C overnight.

5. Add an equal volume of phenol/chloroform/isoamyl alcohol gently down the side of the Falcon tube and place it on it's side to provide the maximum surface area between the DNA solution and the organic phase.
6. After 20 min centrifuge the tube at about 400 r.p.m. for 2 min and remove the organic phase along with most of the interface. Do not attempt to remove all of the interface as this results in an excessive loss of DNA.
7. Repeat steps 5 and 6 using chloroform/isoamyl alcohol. However, this time recover most of the aqueous layer containing the DNA using a 10 ml sterile disposable pipette with an opening of 2–3 mm. To minimize shear, this should be done at a slow rate of about 1 ml/min. Transfer most of the aqueous layer containing the DNA (which may be cloudy at this stage) to a UH 100/75 ultra thimble[a] attached to the holder of a vacuum dialysis apparatus.
8. Concentrate the DNA solution to approximately 5 ml using vacuum dialysis.
9. Dialyse the DNA in the same ultra thimble overnight against 1 litre of TE, 10 mM EDTA at 4°C. This can be done by clamping the ultra thimble and holder assembly (without the suction vessel) such that most of the ultra thimble is immersed in TE contained in a 1 litre beaker.

[a] Ultra thimbles are supplied immersed in alcohol with a sponge insert. Carefully remove the sponge insert, rinse the ultra thimble two or three times in TE, and leave immersed in TE in a 50 ml tube for approximately 1 h. Assemble it into the vacuum dialysis apparatus as described by the manufacturer and apply vacuum to ensure that there are no visible leaks.

Protocol 6. Ligation of insert and vector DNA

Equipment and reagents

- Vacuum dialysis apparatus (see *Protocol 5*)
- Cold water tap aspirator vacuum (see *Protocol 5*)
- 14 ml sterile tubes (see *Protocol 2*)
- T4 DNA ligase (see *Protocol 1*)
- TE (see *Protocol 1*)
- Chloroform/isoamyl alcohol (see *Protocol 1*)
- Phenol/chloroform/isoamyl alcohol (see *Protocol 1*)

Method

1. Add 100 μg dephosphorylated vector arms (see *Protocol 1*) to the DNA solution in the ultra thimble from *Protocol 5*, step 9 (more than a 50-fold molar excess of vector to insert). Mix gently and concentrate by vacuum dialysis to approximately 2 ml.

Protocol 6. *Continued*

2. Dialyse against 1 litre TE for 4–5 h to remove excess EDTA (see *Protocol 5*, step 9). For convenience, this dialysis step may be carried out overnight.
3. Using a 1000 μl pipette tip with the end cut-off to provide an opening of 2–3 mm, transfer the DNA solution into a 14 ml sterile tube. Store the ultra thimble at 4°C for use in step 7 below. (During steps 1–3, the vector is fully mixed with the genomic insert DNA.)
4. Add 10 × ligation buffer (manufacturer's recommended buffer) to achieve a final concentration of 1 × buffer. Mix gently and allow to equilibrate on ice for 1 h.
5. Add 50–100 U ligase in 500–1000 μl of 1 × ligation buffer with gentle mixing. Allow to equilibrate on ice for 1–2 h. Incubate at 12°C overnight.
6. Following ligation the DNA can be extracted once with phenol/chloroform/isoamyl alcohol and once with chloroform/isoamyl alcohol (*Protocol 5*, steps 5–7) to improve transformation efficiencies. However, to minimize shear during these extractions, this step may be omitted.[a]
7. Transfer DNA from step 5 or 6 into the ultra thimble from step 3 above. Dialyse against 1 litre TE for 4–20 h and then concentrate to about 2 ml using vacuum dialysis.
8. Transfer this ligated DNA to two or three sterile microcentrifuge tubes and store at 4°C.

[a] If step 6 is followed, this ligated DNA is stable for several months without loss of transformation efficiency. Long-term stability of ligated DNA prepared by omitting step 6 has not been tested.

One of the major problems with the YAC cloning system has been the high frequency (up to 50%) of chimeras in many of the libraries (2, 24). This is probably due to a combination of non-contiguous insert ligation prior to its ligation to both the vector arms as well as post-transformation recombination in *S. cerevisiae*. In order to overcome some of these problems, in *Protocol 4* above, the DNA plugs in the sample loading slot are recovered along with the DNA fragments in the compression zone. The reasoning behind this is:

(a) This very high molecular weight DNA in the plugs acts as a carrier and protects the clonable insert fragments by increasing the viscosity of the total DNA in solution.

(b) In addition to increasing the number of intact clonable insert fragments, this reduction in DNA damage should reduce the potential number of re-

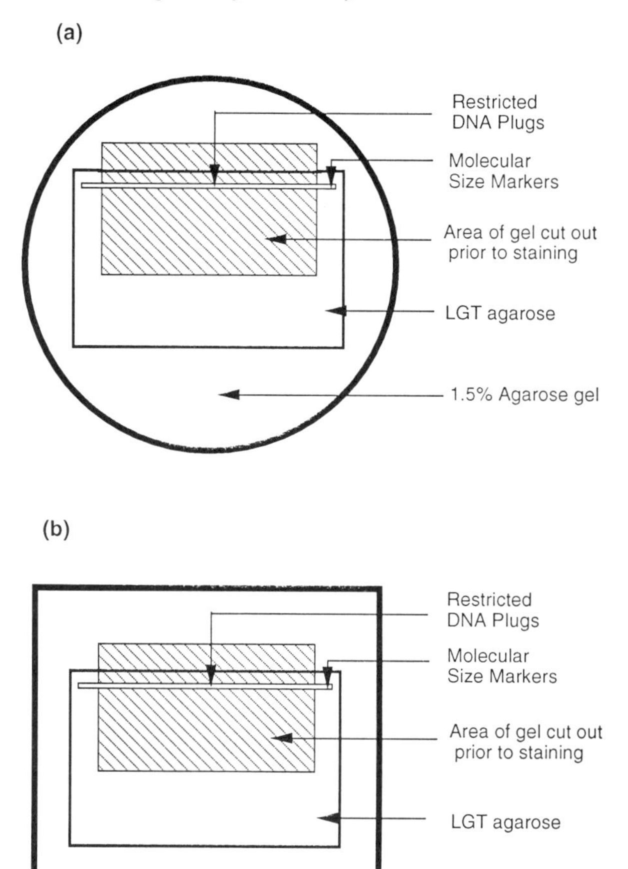

Figure 3. Schematic drawings to illustrate composite agarose gels for preparative PFGE fractionation. A 1.5% agarose gel on the outside provides physical support for the LGT agarose fractionating gel. (a) and (b) represent the circular and square/rectangular gels used in the different types of PFGE apparatus. The shaded area represents the portion of the gel which should be cut out prior to staining the gel with ethidium bromide to visualize the extent of electrophoretic fractionation and the position of the DNA compression zone (also see *Figure 4*).

combinogenic ends following transformation into yeast and thus reduce the number of artefactual (chimeric) YAC inserts.

(c) Inclusion of this very high molecular weight DNA also reduces the probability of a non-contiguous insert ligation resulting in a clonable size insert, thereby reducing the probability of chimeric YACs.

Though the above points have not been proven, a YAC library constructed using the above protocols (11) resulted in a low frequency (approximately

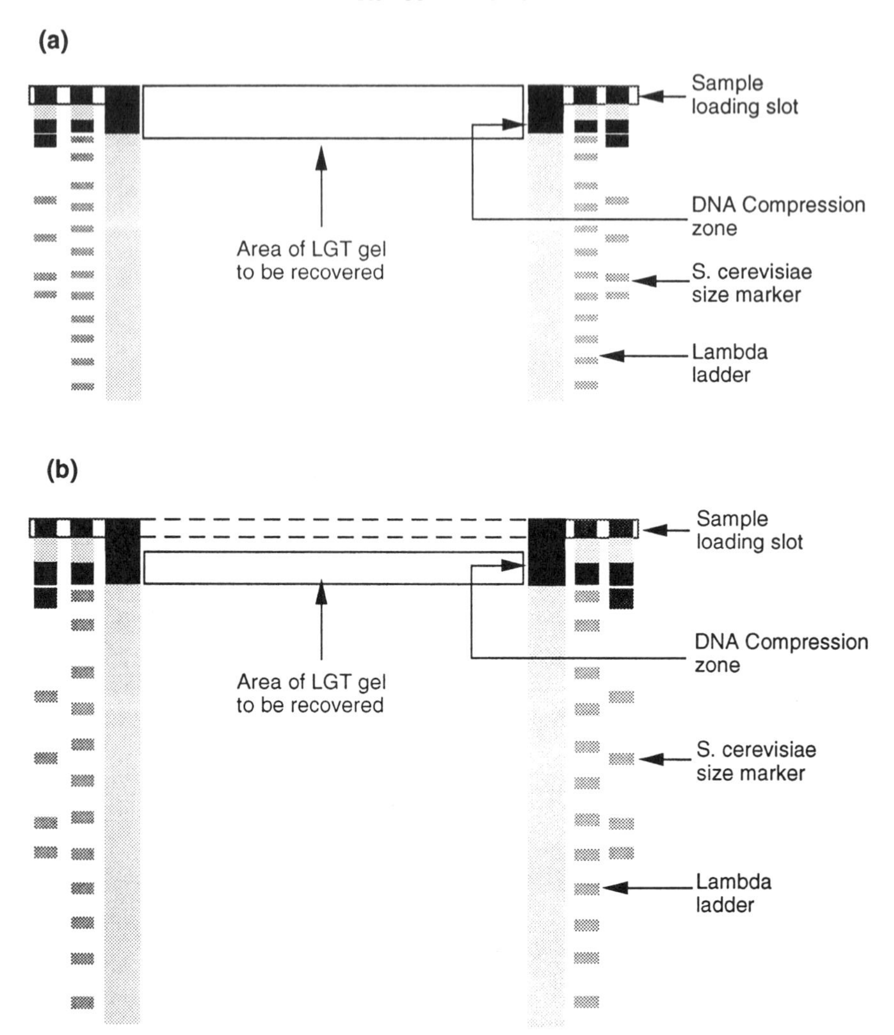

Figure 4. Schematic representation of ethidium bromide stained PFGE gels. (a) Short and (b) long fractionation runs with conditions optimized to select DNA fragments of about 600 kb and larger in the DNA compression zone. In a short run (a) the DNA compression zone is in close proximity to the loading slot and both are recovered together as marked by the box. In a longer run (b) there is a clear separation between the loading slot and the DNA compression zone. In this situation, only the DNA compression zone is recovered as shown by the box.

10%) of chimeric YAC inserts (16 chimeras detected following analysis of 159 independent YAC clones). Another approach is to use recombination deficient host strains. Studies on YAC stability in different strains of *S. cerevisiae* (25) and the construction of a mouse genomic YAC library in a *rad52* deficient strain support this hypothesis (26).

Two modifications of the above protocols are also worth noting. First is the use of only the DNA compression zone (without the plugs in the loading well) as the source of PFGE fractionated insert DNA (13, 14). If that is the aim, the run time in *Protocol 4*, step 3 should be increased to clearly resolve the compression zone away from the sample loading slot (see *Figure 4b*). The second variation is the use of a second size fractionation step following ligation of the vector arms to the insert DNA (12–14). This helps to remove the unligated vector arms and re-selects ligated DNA in favour of YACs above a predetermined size.

4. YAC library construction

The complexity of a YAC library is determined by the total genomic size of the organism providing the insert DNA (e.g. 3×10^6 kb for a haploid human genome), the average YAC insert size, and the number of recombinant clones making up the library. A complexity of three to four genome equivalents, providing approximately 95% chance of finding a given sequence, is adequate for most applications. However, libraries constructed with the aim of mapping the entire human genome are generally more complex (approximately six to ten genome equivalents) in order to increase the chances of finding clones corresponding to underrepresented regions. This also compensates for cloning artefacts, such as chimerism, rearrangements, and deletions, which can cause problems with random fingerprinting approaches aimed at mapping large regions of complex genomes (2). An alternative to the construction of libraries with increased complexity is to construct two or more YAC libraries using different restriction enzymes. This should reduce the chances of the underrepresentation of regions of the genome particularly rich or poor in restriction sites for the enzymes used to construct the libraries. In order to define an end-point, it is advisable to consider these issues and decide on the final complexity being aimed for before embarking on repeated transformation and storage of recombinant clones. This is perhaps best done after an estimation of the average YAC size has been obtained following the first one or two transformations (section 4.2).

4.1 Transformation

This is based on the method of Burgers and Percival (27). Using a single batch of ligated DNA, transformation efficiencies can vary by as much as one order of magnitude (100 to $>$ 1000 transformants/μg). Increasing the DNA concentrations or addition of extra carrier DNA does not result in a consistent improvement. This variation is probably due to the colonies used to initiate the AB1380 cultures or in the processing steps involved in making competent spheroplasts. Uncut plasmid YCp50 (an 8 kb plasmid containing the *ura3* selectable marker gene) should be used as control DNA to monitor day to day variation in transformation efficiency (10^6–10^7/μg plasmid DNA).

Protocol 7. Transformation

Equipment and reagents

All solutions and media should be made up in glass double-distilled water and sterilized by autoclaving except where mentioned otherwise.

- SD: 0.7% yeast nitrogen base without amino acids, 2% glucose, 5.5 mg/100 ml of adenine and tyrosine, pH 7.0; for single (*-ura*) selection add 7 ml of 20% casamino acids solution and 2 ml of 1% tryptophan/100 ml; for double selection (*-ura, -trp*), add 7 ml of 20% casamino acids/100 ml
- SCE: 1.0 M sorbitol, 0.1 M sodium citrate, 10 mM EDTA, pH 5.8, autoclave, and add β-mercaptoethanol to 28 mM before use
- CaS: 1.0 M sorbitol, 10 mM Tris–HCl pH 7.5, 10 mM $CaCl_2$
- PEG: 20% PEG 8000 (PEG 6000 can also be used), 10 mM $CaCl_2$, 10 mM Tris–HCl pH 7.5 (filter sterilize)
- YPD: 2.0% glucose, 2.0% bacto peptone, 1.0% yeast extract
- Recovery plates: 2.2% bacto agar in SD—to provide double selection, add 7 ml of 20% casamino acids solution/100 ml molten agar at 50–60°C prior to pouring these plates
- SOS: 10 ml 2 M sorbitol, 6.7 ml YPD, 130 μl 1 M $CaCl_2$, 3.17 ml water (all sterilized prior to mixing)
- Transformation plates: 2.2% bacto agar in SD containing 1 M sorbitol—for single selection, add 7 ml of 20% vitamin assay casamino acids (Difco) and 2 ml of 1% tryptophan/100 ml molten agar at 50–60°C prior to pouring these plates
- Top agar for plating out recombinants: 3.3% bacto agar in SD containing 1 M sorbitol—for single selection, add 7 ml of 20% vitamin assay casamino acids and 2 ml of 1% tryptophan/100 ml molten agar at 50–60°C
- YPD plates: 2.0% bacto agar in YPD
- Lyticase (Sigma or similar): make up a stock solution to 5000–10 000 U/ml in sterile 50 mM Tris–HCl pH 7.5, 1 mM EDTA, 10% glycerol—test efficiency of spheroplast formation (see steps 1–6 below and note[a] and store in small aliquots (~ 500 μl) at −20°C

Method

1. Streak out *S. cerevisiae* AB1380 cells on a YPD plate and grow at 30°C for two to three days. Once grown, the plate can be stored at 4°C and colonies used to inoculate cultures for up to three weeks.
2. Inoculate 10 ml YPD culture with one colony and grow overnight at 30°C
3. Inoculate 50 ml YPD culture with 5–10 ml of the overnight culture (to an OD_{600} of 0.1–0.3) and grow at 30°C to an OD_{600} of 0.7–0.8 units.
4. Harvest the cells by centrifugation at 1500 r.p.m for 10 min (J-6B Beckman or its equivalent). Wash the cells in 20 ml sterile distilled water and pellet them by centrifugation at 1500 r.p.m. for 5 min.
5. Wash the cells in 20 ml 1 M sorbitol and pellet them by centrifugation at 1500 r.p.m. for 5 min.
6. Resuspend the cells in 20 ml SCE containing 40 μl β-mercaptoethanol. Add sufficient pre-titrated lyticase to achieve 90% spheroplast formation in approximately 15 min.[a]
7. When 80–90% cells have formed spheroplasts (this takes approximately 15 min) recover the spheroplasts by centrifugation at 1000–1500 r.p.m. for 5 min.

8. Resuspend the spheroplasts gently in 20 ml 1 M sorbitol by first flicking the tube containing the pellet to dislodge the spheroplasts prior to adding sorbitol. Centrifuge at 1000 r.p.m. for 3–4 min to recover the spheroplasts.
9. Wash the spheroplasts once in 20 ml CaS and then resuspend in 2 ml CaS.[b]
10. Using a 200 μl dispensing pipette and a tip with the end cut-off to provide an opening of 2–3 mm, transfer ~ 0.5 μg ligated DNA (10–20 μl) to a 14 ml sterile snap-top tube. Add 350 μl spheroplasts[c], mix by gentle swirling, and leave at room temperature for 10 min.
11. Add 3.5 ml PEG solution at room temperature and mix by inverting the tube two or three times. Leave at room temperature for 10 min and then centrifuge at 1000 r.p.m. for 5 min. Resuspend the cells by first flicking the tube and then adding 500 μl SOS. Incubate at 30°C for 20–40 min.
12. Add 6 ml top agar at 45°C, mix quickly, and plate out on a single selection (-*ura*) plate[c] which has been pre-warmed to 37°C. Allow to set and incubate at 30°C. Transformants can be seen after three to five days.

[a] This is an important step and may require some trial and error. Incubate cells and lyticase (1000–2000 U) at 30°C with occasional mixing. Monitor spheroplast formation every 5 min by mixing a small aliquot of cells with 10% SDS on a microscope slide and examine under a phase contrast microscope. Spheroplasts appear as dark ghosts compared to bright and shiny intact yeast cells. It should take approximately 15 min to get to 80–90% spheroplasts. If it takes more than 20 min to get to this stage, more lyticase should have been used. In this situation, it is advisable to start again from step 2 or 3 since cells which take more than 20 min to get to about 90% spheroplasts may result in reduced transformation efficiencies. An alternative method for monitoring spheroplast formation is to follow the decrease in OD_{800} of a tenfold dilution of spheroplasts in water (27).
[b] It has been reported that inclusion of polyamines during the transformation steps protects large DNA from damage by shearing and consequently helps in achieving a large average insert size (28). Polyamines can be used at a concentration of 0.75 mM spermidine and 0.3 mM spermine, but they must be present in all solutions used in steps 9–11.
[c] Using the above protocol, sufficient spheroplasts are produced to allow the use of up to five tubes containing DNA in step 10 above. The number of DNA containing tubes used in any single transformation experiment will be determined by the requirements and experience of the operator.

4.2 Library characterization

It may be prudent to do some initial characterization of the YAC clones derived after the first one or two transformations before proceeding with the construction of a complete library. This involves estimation of the average YAC size, and if possible, some estimate of the integrity of the YACs, i.e. frequency of chimerism (2, 24). The average size can be estimated by preparing individual plugs from about 100 recombinants and sizing them by

PFGE. Alternatively, an estimate of the average size can be obtained by using plugs containing DNA of 96 clones (11) (see *Protocol 8B*). If five to ten such plugs are run on a PFGE gel and a Southern blot of this gel is hybridized with a pBR322 probe, all YACs will be detected (11). This should provide a reasonable estimate of the range of YAC sizes in the library. Testing the integrity (i.e. if the YACs contain contiguous DNA fragments) requires either fluorescence *in situ* hybridization (FISH) analysis (29, 30) or the PCR amplification of YAC end-specific sequences (31, 32) from mono-chromosome hybrid panels. These techniques are outside the scope of this chapter and the reader should consult relevant published literature (29–32).

4.3 Yeast chromosomal DNA mini-preps

DNA prepared by embedding YAC clone cells in agarose can be used for a variety of diverse applications including YAC sizing, restriction digests to construct physical maps of YACs, PCR screening, isolation of YAC ends, FISH analysis, construction of YAC sublibraries, and preparative PFGE to obtain individual YAC DNA free from the background yeast DNA.

Protocol 8. YAC DNA plugs[a]

Equipment and reagents

- YLB: 1.0% lithium dodecyl sulfate, 100 mM EDTA (added from a stock of 0.5 M EDTA pH 8.0), 10 mM Tris–HCl pH 8.0
- YRB: 1.0 M sorbitol, 10 mM Tris–HCl pH 7.5, 20 mM EDTA (added from a stock of 0.5 M EDTA pH 8.0)—autoclave and add β-mercaptoethanol to 14 mM before use
- SD (see *Protocol 7*)
- EDTA (see *Protocol 1*)
- Lyticase (see *Protocol 7*)
- LGT agarose (see *Protocol 2*)
- Plug mould (see *Protocol 2* and *Figure 2*)
- Recovery plates (see *Protocol 7*)

A. *DNA plugs from individual YAC clones*

1. Streak out the YAC clone of choice on a recovery plate and grow for two to three days at 30°C.
2. Inoculate a 10 ml SD (double selection) culture with a single colony and grow about 20 h at 30°C.
3. Harvest the cells by centrifugation, wash once in 50 mM EDTA pH 8.0, and resuspend in 500 μl YRB containing 50 U/ml lyticase.
4. Incubate at 37°C and monitor spheroplast formation (see footnote[a], *Protocol 7*) until 80–90% of the cells form spheroplasts.
5. Gently mix the cells 1:1 (v/v) with a 1% solution of LGT agarose in YRB at 37°C and transfer to a plug mould (see *Protocol 2*). Allow to set.
6. Transfer the plugs into 4–5 vol. of YLB and gently shake at room temperature for 1–2 h. Replace the solution with 10 vol. of YLB and incubate at 40–50°C overnight with gentle shaking (see *Protocol 2*). Rinse the plugs once with YLB for 1 h and store in YLB at room

temperature. This DNA is essentially intact and stable by PFGE analysis for a year or longer.

B. *Recovery plate DNA plugs*

1. Following transfer of the cells from recovery plates to microtitre plates (*Protocol 9, step 3*), or inoculation of fresh recovery plates with 96 recombinant clones from each microtitre plate (see footnote[a], *Protocol 9*), incubate the recovery plate at 30°C overnight.
2. Pour 10 ml 500 mM EDTA on the top of each plate and use a bent glass rod to mix all the cells. Pour the cells into a 50 ml Falcon tube. Rinse the plate with an additional 10 ml EDTA to recover most of the remaining cells. Add these to the rest in the Falcon tube.
3. Make DNA plugs using part A, steps 3–6 (above). These plugs contain chromosomal DNA from 96 recombinant clones.

[a] See also Chapter 7, *Protocol 2*.

4.4 Gridding and storage

For screening and easy access, recombinant clones are most conveniently stored in 96-well microtitre plates. It is now possible to use robotics ('The colony picker' from Hybaid or similar) to pick individual clones from the transformation plates and format them into microtitre plates. However, since this may not be a cost-effective option for laboratories constructing a single YAC library, a manual alternative is described in *Protocol 9*.

All storage should be at −70°C. For ease of access, at least one copy of the library should be stored as individual clone cultures in double selection SD (see *Protocol 7*) containing 20–25% glycerol in microtitre plates. Storage of additional back-up copies depends on freezer space and resource constraints. To protect against contamination, accidents, and freezer failures, it is advisable to store at least one back-up copy of the entire library in another −70°C freezer (preferably on a separate electrical circuit to the freezer housing the working copy).

Protocol 9. Gridding and storage

Equipment and reagents

- Sterile microtitre plates with lids
- SD (see *Protocol 7*)
- SD, 20% glycerol (autoclave to sterilize)
- Sterile plastic tissue culture loops
- Recovery plates with a microtitre plate 96 position template attached to the bottom to provide orientation for each of the 96 wells in a microtitre plate
- Wooden cocktail sticks (catering suppliers): autoclave in glass beakers covered with aluminium foil finely pierced on top
- Wellrep 1: a 96 prong microtitre well replication device (Cat. No. WR081, Denley Instruments Ltd., Billinghurst, West Sussex RH14 9EY)

Protocol 9. *Continued*

Method

1. Use sterile cocktail sticks to pick individual clones from a single selection primary transformation plate (see *Protocol 7*). Streak out each clone on to one of the 96 positions on a double selection (*-ura, -trp*) recovery plate.
2. Incubate recovery plates at 30°C over two to three days so that there are large numbers of cells for each clone. This also allows for clear expression of the red colour in the recombinant clones.
3. Use a sterile microtitre plate well replicator (Wellrep 1) to simultaneously transfer all 96 clones[a] from a recovery plate to the 96 wells of a microtitre plates containing 100 μl double selection SD, glycerol. This is best done by first dipping the well replicator in a plate containing SD so that each of it's pins pick up approximately 10 μl of media. Align and place the well replicator on to the recovery plate clones (the media helps to moisten the cells), move it gently in a small circular motion to ensure that each pin picks up approximately 50% of the yeast cells. Transfer the cells to the microtitre plate containing SD, glycerol.
4. Rinse the well replicator pins twice in water and once in absolute alcohol. Flame sterilize and cool before reuse in step 3 above.

[a] To ensure that each recovery plate in step 3 and the corresponding microtitre plate contains cells from 96 recombinant clones, all non-recombinant (white) clones can be replaced by recombinant (red) clones from a back-up recovery plate. This is done by using a sterile plastic tissue culture loop to remove most of the non-recombinant cells from any of the 96 positions and replacing with recombinant clone cells from the back-up recovery plate. It should not be necessary to do this for more than four or five clones per recovery plate. If the frequency of non recombinants is > 5%, it may be too tedious to replace individual clones with recombinant ones. In this case, an alternative to step 3 is to individually transfer 96 recombinant clones from the recovery and back-up plates to the microtitre plate and then use Wellrep 1 to inoculate a fresh recovery plate for making DNA plugs from pools of 96 recombinant clones (*Protocol 8*).

5. Screening YAC libraries

YAC libraries stored in microtitre plates are amenable to screening protocols based on hybridization or the polymerase chain reaction (PCR). Essentially, hybridization protocols are based on robotic assisted generation of high density arrays of clones on nylon filters. Several different arrays have been described (31, 33, 34) ranging from several hundred clones to approximately 20000 clones on a 22 × 22 cm filter. The density and format of the array is generally a blend between personal preference and the capability of the available robotics facility. Details on the generation of high density clone arrays and their use are clearly outside the scope of this chapter and the reader is referred to the relevant references (31, 33, 34).

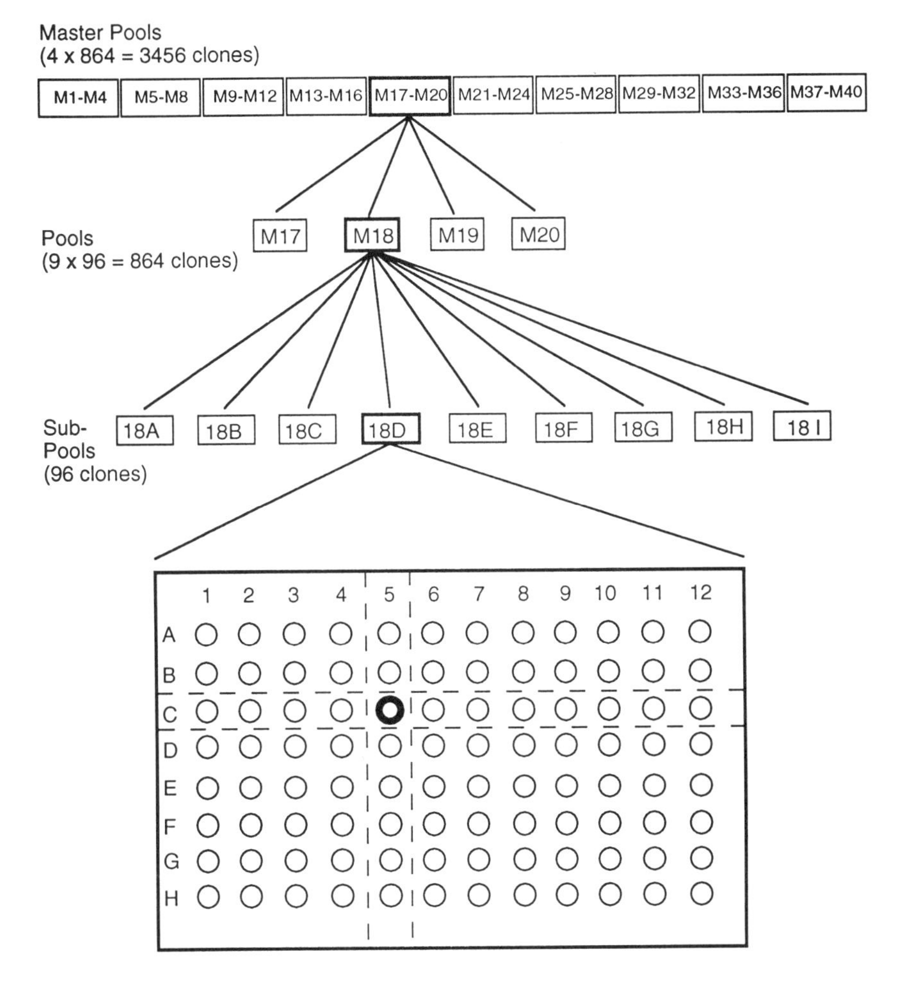

Figure 5. PCR-based YAC library screening: one possible scheme for making pools of YAC clone DNA from a library consisting of 34 560 clones in 360 microtitre plates. Subpools contain DNA from all 96 clones in a single microtitre plate (see *Protocols 8B* and *10*). Pools M1–M40, each contains DNA from nine subpools (9 × 96 = 864 clones). The ten master pools are each made up of DNA from four pools (M1–M40) and therefore contain DNA from 864 × 4 = 3456 clones. In this example, the first level of PCR screen identifies master pool 'M17–M20'. The second level uses these four pools and identifies 'M18' as the positive pool. The third level identifies '18D' as the positive subpool. Following identification of the microtitre plate(s) containing the positive clone(s), the final round of screening is done using pools of clones from the eight rows (A–H) and 12 columns (1–12) of the individual microtitre plate (see *Protocol 10*). The positive clone is identified by the coordinates of the positive row 'C' and column '5'. In this example, the positive clone is 18DC5.

PCR screening is the method of choice for most laboratories which are not equipped with robotics to generate high density clone arrays. This relies on the use of a pair of oligonucleotides designed to selectively amplify by PCR, a specific region of the human or other genome under investigation, e.g. a sequence tag site (STS) (35). Pools of YAC clones are used as the DNA substrate such that with every round of PCR screening, the number of clones in a pool decreases. The eventual aim is to identify all positive clones within a library. There are several formats for screening (31, 36) and generally the format of choice is a compromise between the number of PCR reactions and the DNA pooling strategies. One possible scheme is shown in *Figure 5*.

Protocol 10. Making pools of YAC clone DNA for PCR screening

Equipment and reagents

- TE (see *Protocol 1*)
- Sterile polypropylene tubes (see *Protocol 2*)
- Wellrep 1 (see *Protocol 9*)
- Sterile water (see *Protocol 2*)

A. *Making master and subpools*

1. Add one DNA plug from each of a specified number (9 in *Figure 5*) of recovery plate DNA plugs (see *Protocol 8B*) into 50 ml tubes containing 40 ml TE. Wash with gentle mixing (on a tube roller or orbital shaker) for 2–4 h.
2. Repeat washing with approximately 40 ml TE, once overnight, and again once over about 2 h.
3. Drain out the TE and incubate the DNA plugs at 65–68°C for 10–20 min to melt the LGT agarose.
4. Allow to cool to about 40°C and mix thoroughly (vortex) whilst maintaining a temperature of greater than 35°C to ensure that the agarose remains in solution. This mixture of DNA forms the pools (M1–M40 in *Figure 5*).
5. Aliquot into 1.5 ml microcentrifuge tubes (100 μl each) and store at room temperature. For storage over several years, −20°C may be preferable.
6. For use in PCR screening, take one of the 100 μl aliquots and add 900 μl TE. Heat to 65–68°C to melt the agarose and mix well. Use 1–5 μl as the DNA substrate for amplification by PCR.
7. To make master pools, take the 100 μl aliquots (step 5 above) of four pools (*Figure 5*), melt, and vortex mix them together. Aliquot into 100 μl portions and process as in step 6 above.
8. To make subpool DNA, process individual recovery plate DNA plugs (see *Protocol 8B*) as in steps 2–6 above using 14 ml sterile tubes instead of 50 ml ones.

B. *DNA pools of microtitre plate rows and columns*

1. Use Wellrep 1 to simultaneously inoculate the 96 clones from microtitre plates identified as ones containing positive clones by PCR screening (*Figure 5*) on to two recovery plates (*Protocol 9*).
2. Incubate the plates at 30°C for 24–48 h (until all clones have grown well).
3. Use one plate to transfer cells of clones from each of the rows (A1–A12, B1–B12, etc.) into 1.5 ml microcentrifuge tubes containing 1 ml distilled water.
4. Use the second plate to transfer clones from each of the columns (A1–H1, A2–H2, etc.) into 1.5 ml microcentrifuge tubes containing 1 ml distilled water.
5. Vortex mix all tubes and then pellet the cells. There is sufficient DNA released during this step to provide an adequate substrate for PCR amplification.[a]
6. 1–5 μl of the supernatant from these pools of rows and columns can be used as the DNA substrate for amplification by PCR. The individual positive clone is identified by the coordinates of the positive row and column (*Figure 5*).

[a] These pools of rows and columns can be stored for approximately one year for subsequent use in other screens where positive clones are localized to the same microtitre plate.

6. Conclusions

It should be clear by now that the construction of a primary YAC library requires a substantial investment in time and resource. Before embarking on this exercise, the possibility of using one of the existing YAC libraries should be given serious consideration. If the decision is to proceed with the construction of a library, it is essential to carefully plan all the steps involved. It may also be worth considering potential future use as this may impact on the choice of vector, average insert size, and the final complexity of the YAC library.

References

1. Burke, D. T., Carle, G. F., and Olson, M. V. (1987). *Science*, **236**, 806.
2. Cohen, D., Chumakov, I., and Weissenbach, J. (1993). *Nature*, **366**, 698.
3. Shero, J. H., McCormick, M. K., Antonarakis, S. E., and Hieter, P. (1991). *Genomics*, **10**, 505.
4. Pavan, W. J., Hieter, P., and Reeves, R. H. (1990). *Proc. Natl Acad. Sci. USA*, **87**, 1300.
5. Pavan, W. J., Hieter, P., Griffin, C. A., Hawkins, A. L., and Reeves, R. H.

(1992). In *Techniques for the analysis of complex genomes* (ed. R. Anand), pp. 173–96. Academic Press Ltd., London.
6. Smith, D. R., Smyth, A. P., and Moir, D. T. (1990). *Proc. Natl Acad. Sci. USA*, **87**, 8242.
7. Sambrook, J., Fritsch, E. F., and Maniatis, T. (ed.) (1989). *Molecular cloning, a laboratory manual* (2nd edn). Cold Spring Harbor Laboratory Press, Cold Spring Harbor, New York.
8. Coulson, A., Waterston, R., Kiff, J., Sulston, J., and Kohara, Y. (1988). *Nature*, **335**, 184.
9. Brownstein, B. H., Silverman, G. A., Little, R. D., Burke, D. T., Korsmeyer, S. J., Schlessinger, D., *et al.* (1989). *Science*, **244**, 1348.
10. Traver, C. N., Klapholz, S., Hyman, R. W., and Davis, R. W. (1989). *Proc. Natl Acad. Sci. USA*, **86**, 5898.
11. Anand, R., Riley, J. H., Butler, R., Smith, J. C., and Markham, A. F. (1990). *Nucleic Acids Res.*, **18**, 1951.
12. Imai, T. and Olson, M. V. (1990). *Genomics*, **8**, 297.
13. Albertsen, H. M., Abderrahim, H., Cann, H. M., Dausset, J., Le Paslier, D., and Cohen, D. (1990). *Proc. Natl Acad. Sci. USA*, **87**, 4256.
14. Larin, Z., Monaco, A. P., and Lehrach, H. (1991). *Proc. Natl Acad. Sci. USA*, **88**, 4123.
15. Edwards, K. J., Thompson, H., Edwards, D., De Saizieu, A., Sparks, C., Thompson, J. A., *et al.* (1992). *Plant Mol. Biol.*, **19**, 299.
16. Guzman, P. and Ecker, J. R. (1988). *Nucleic Acids Res.*, **16**, 11091.
17. Garza, D., Ajioka, J. W., Burke, D. T., and Hartl, D. L. (1989). *Science*, **246**, 641.
18. Anand, R., Villasante, A., and Tyler-Smith, C. (1989). *Nucleic Acids Res.*, **17**, 3425.
19. Libert, F., Lefort, A., Okimoto, R., Womack, J., and Georges, M. (1993). *Genomics*, **18**, 270.
20. Elvin, P., Slynn, G., Black, D., Graham, A., Butler, R., Riley, J., *et al.* (1990). *Nucleic Acids Res.*, **18**, 3913.
21. Huxley, C., Hagino, Y., Schlessinger, D., and Olson, M. (1991). *Genomics*, **9**, 742.
22. Capecchi, M. R. (1993). *Nature*, **362**, 205.
23. Southern, E. M., Anand, R., Brown, W. R. A., and Fletcher, D. S. (1987). *Nucleic Acids Res.*, **15**, 5925.
24. Green, E. D., Riethman, H. C., Dutchik, J. E., and Olson, M. V. (1991). *Genomics*, **11**, 658.
25. Vilageliu, L. and Tyler-Smith, C. (1992). In *Techniques for the analysis of complex genomes* (ed. R. Anand), pp. 93–112. Academic Press Ltd., London.
26. Chartier, F. L., Keer, J. T., Sutcliffe, M. J., Henriques, D. A., Mileham, P., and Brown, S. D. M. (1992). *Nature Genetics*, **1**, 132.
27. Burgers, P. M. J. and Percival, K. J. (1987). *Anal. Biochem.*, **163**, 391.
28. Connelly, C., McCormick, M. K., Shero, J., and Hieter, P. (1991). *Genomics*, **10**, 10.
29. Kievitz, T., Devilee, P., Wiegant, J., Wapenaar, M. C., Cornelisse, C. J., Van Ommen, G. J. B., *et al.* (1990). *Cytometry*, **11**, 105.
30. Gingrich, J. C., Shadravan, F., and Lowry, S. R. (1993). *Genomics*, **17**, 98.

31. Bentley, D. R. (1992). In *Techniques for the analysis of complex genomes* (ed. R. Anand), pp. 113–35. Academic Press Ltd., London.
32. Silverman, G. A. (1993). *PCR Methods Applications*, **3**, 141.
33. Bentley, D. R., Todd, C., Collins, J., Holland, J., Dunham, I., Hassock, S., *et al.* (1991). *Genomics*, **12**, 534.
34. Ross, M. T., Hoheisel, J. D., Monaco, A. P., Larin, Z., Zehetner, G., and Lehrach, H. (1992). In *Techniques for the analysis of complex genomes* (ed. R. Anand), pp. 137–53. Academic Press Ltd., London.
35. Olson, M., Hood, L., Cantor, C., and Botstein, D. (1989). *Science*, **245**, 1434.
36. Jones, M. H., Khwaja, O. S. A., Briggs, H., Lambson, B., Davey, P. M., Chalmers, J., *et al.* (1994). *Genomics*, **24**, 266.

5

Amplification of DNA microdissected from mitotic and polytene chromosomes

ROBERT D. C. SAUNDERS

1. Introduction

Techniques permitting access to defined chromosomal regions are invaluable for the analysis of complex genomes. Microcloning is one such technique, which is particularly powerful when applied to dipteran polytene chromosomes. Microcloning was first described by Scalenghe *et al.* (1), who microdissected approximately 100 kb from the *white* region of the *Drosophila melanogaster* X chromosome, and recovered DNA fragments cloned in a phage lambda insertion vector, essentially by conventional cloning methods in miniature. Despite the obvious technical problems in transferring this technique to the study of mitotic chromosomes, microcloning has been applied to the analysis of specific chromosome regions of a variety of organisms.

The application of the polymerase chain reaction to this methodology was described by Lüdecke *et al.* (2), for the amplification of DNA microdissected from human chromosomes, prior to its recovery by cloning. The method devised by Saunders *et al.* (3) and Johnson (4) has greatly simplified this procedure, by using cohesive-ended oligonucleotide adapters rather than blunt-ended plasmid DNA to provide priming sites, and by the use of thermostable DNA polymerase.

Perhaps the major use of DNA amplified in this way is as a probe for screening libraries. This is being used in one of the *Drosophila* genome mapping projects (5, 6). A similar approach has been used to create a low-resolution map of the *Anopheles gambiae* genome (7), and here the pools of amplified DNA are being cloned and used to facilitate cloning *Anopheles* homologues of genes cloned from *Drosophila* and other species. Microdissection is a powerful means by which STS, microsatellite, and RFLP markers can easily be obtained. These and other applications of chromosome microdissection to the analysis of genomes and specific loci are discussed in section 4.

5' 3'

pGATCAGAAGCTTGAATTCGAGCAG
TCTTCGAACTTAAGCTCGTC

3' 5'

Figure 1. The double-stranded oligonucleotide adapter.

The basic approach is to excise fragments from fixed chromosome preparations adhering to the underside of a coverslip in an oil filled chamber. All the micromanipulation steps are carried out using microinstruments pulled from glass capillaries, mounted on a micromanipulator (see section 2.2.1). The glass microneedle used for the microdissection is also used to transfer the chromosome fragment to a microdrop of buffer. Sequential additions of buffers and enzymes to this microdrop allow the required nucleic acid manipulations to be carried out under the microscope. Briefly, the extraction droplet is incubated to release the DNA from the chromosome fragment, then extracted with phenol to remove traces of proteinase. The DNA is then digested with a restriction enzyme to reduce the mean fragment length to a manageable size. To provide priming sites for PCR amplification, double-stranded oligonucleotide adapters bearing compatible cohesive ends are ligated to the DNA fragments. Finally, the microdrop is removed from the oil chamber, for PCR amplification.

The adapter consists of a 20-mer annealed to a 24-mer (*Figure 1*) resulting in a double-stranded oligonucleotide with one cohesive *Sau*3AI terminus, which is phosphorylated to permit double-stranded ligation to the *Sau*3AI digested chromosomal DNA. The primer used in the amplification of the chromosomal DNA is the 20-mer strand of the adapter. The arrangement of apparatus described in this chapter is not the only possibility. Alternative arrangements include stage mounted micromanipulators, inverted microscopes, and the use of lasers (8) or glass knives (9) for microdissection. The system used will largely depend on which items of equipment are already present in the laboratory. Moreover there are other possibilities with respect to the amplification of microdissected material. For example some investigators have primed PCR amplifications directly on microdissected polytene chromosomal DNA, though this tends to yield products of low sequence complexity (10).

2. Materials

2.1 Equipment

The oil chamber in which all the biochemical steps are carried out is a thick glass slide (35 mm × 70 mm × 6 mm thick) with a groove cut in it (*Figure 2*). The groove should be about 3.5 mm deep. Coverslips are placed so as to

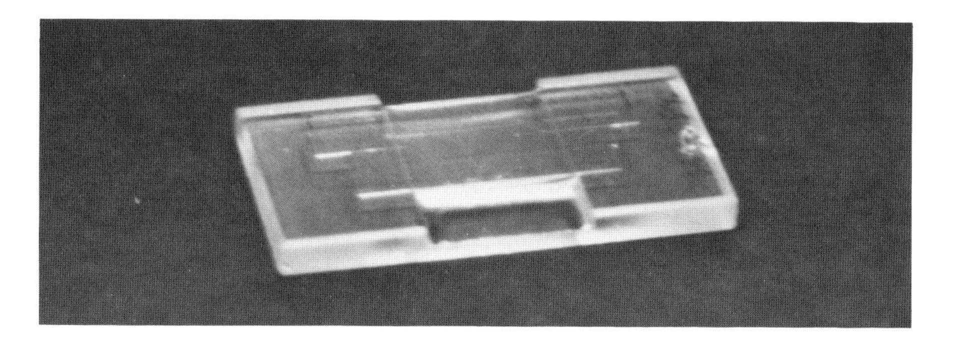

Figure 2. The glass oil chamber. The oil chamber is constructed by cutting a groove in a thick glass slide. The three narrow coverslips can be seen bridging the groove. The space between coverslips and groove is filled with buffer-saturated paraffin oil.

bridge this groove, and the space filled with paraffin oil (IR spectroscopy grade) saturated with R buffer (10 mM Tris–HCl pH 7.5, 100 mM NaCl, 10 mM $MgCl_2$, 10 mM 2-mercaptoethanol). Suitable coverslips are made by cutting 20 mm × 40 mm coverslips with a diamond pencil, to yield 5 mm × 40 mm coverslips. Three of these coverslips are used at any one time. Some of these coverslips should be siliconized, and some left unsiliconized. The reaction microdrops are positioned on the underside of the coverslips and therefore surrounded by oil, to avoid evaporation. Reaction components are added to the reaction drops using micropipettes operated by a free standing micromanipulator. The preferred micromanipulator, such as the de Fonbrune model, are operated by a remote joystick, minimizing transmission of vibrations to the microinstruments. The microinstruments enter the chamber from the rear, and are mounted horizontally. It follows therefore that the microscope must enable focusing by a moving objective rather than a moving stage. A compound microscope is required, with × 10 and × 40 phase objectives which do not require immersion oil. It must be stressed that other arrangements of apparatus can be used. In particular, other designs of micromanipulator may permit the use of an inverted microscope. The arrangement of apparatus is shown in *Figure 3*. To produce the micropipettes and microneedles used in these protocols, a suitable microforge is required, such as the model from de Fonbrune, or the Narashige MF-9. These instruments are essentially a microscope with an element attached to the objective. The angle and position of the glass capillary is finely adjustable, as is the element. The element temperature is controllable. In extremis, such a device can be constructed using a dissecting microscope.

Liquid is picked up and expelled by micropipettes by means of a 10 ml syringe connected to the micropipette by a three-way valve, which allows the air in the system to be expelled without discharging the reagent. To prevent

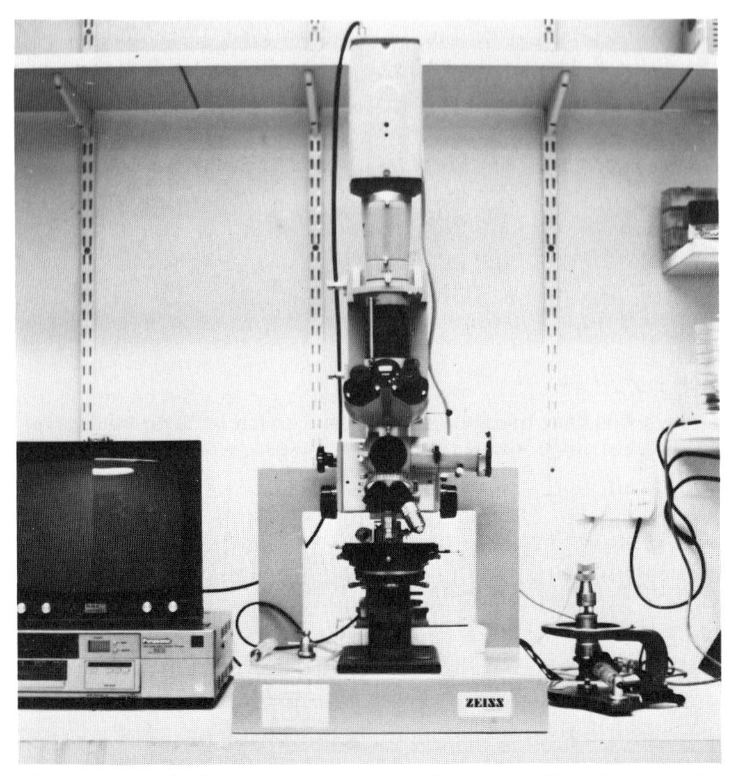

Figure 3. Microscope and micromanipulator set-up for a microdissection experiment. The microscope has a fixed stage on which the oil chamber is mounted, with the micromanipulator located to the rear. The joystick controlling the micromanipulator is to the right, and the three-way valve attached to the micropipette is on the left. This system comprises a Zeiss ACM microscope with dry phase contrast × 10 and × 40 objectives, and a de Fonbrune micromanipulator (Micro-Instruments (Oxford) Ltd, Oxford). In this set-up, the image is displayed upon a video monitor.

evaporation and denaturation of enzymes at the air/buffer interface, the micropipettes are partly filled with oil prior to drawing up reagents (with the exception of chloroform, see *Protocol 4*). This also has the effect of damping the action of the syringe. Always mark the reaction coverslip so its orientation can be recognized, and note the stage coordinates of each reaction drop dispensed. It is also a good idea to dispense a row of reaction droplets, even if only one is to be used, as this makes it easy to find them.

2.2 Construction of microinstruments

2.2.1 Microneedles

Microneedles are prepared from 200 μl microcapillaries (*Figure 4D–G*). The capillaries should be drawn-out over a Bunsen flame in two stages to about

0.5 mm diameter. This is broken, leaving about 5–8 cm of pulled capillary. The end of the drawn part of the capillary is melted with the microforge filament which is then drawn away and the heat reduced to form the point. With a little practice, this can produce good needle points. The capillary is then rotated through 180°, and heat applied to change the bend angle from 90° to approximately 135°. Note that the length of the upturn should be short enough that the needle can enter the oil chamber without obstruction. The best way to assess the quality of needles is to try them out on chromosome preparations. They should be able to cut out single bands of polytene chromosomes (or portions of the arms of mitotic chromosomes).

2.2.2 Micropipettes

Micropipettes are also made from 200 μl capillaries (*Figure 4A–C*). First the capillary is pulled as for the needle. The capillary is positioned by the microforge filament so that it is at about 45° to the filament, and clamped so that it is over the filament. A small piece of Blu-tac or similar adhesive putty is attached to the end of the capillary as a weight. The filament is heated, and the bottom end is allowed to drop to vertical. On further heating, this end

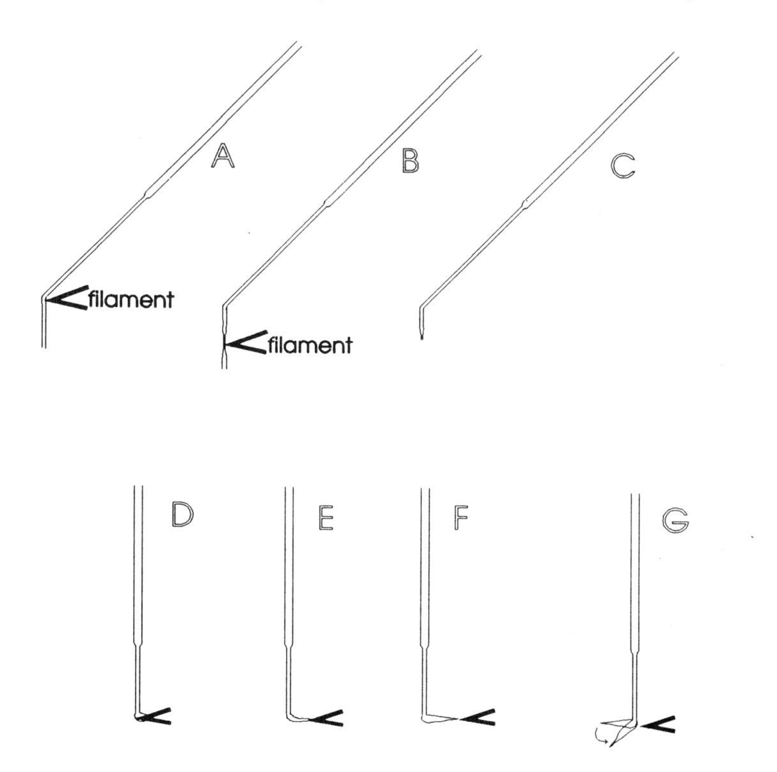

Figure 4. Steps in the construction of micropipettes (A–C) and microneedles (D–G).

will drop, pulling the capillary to a fine tube, which should be broken apart with forceps. As with the microneedle, the upturn should be short enough to allow the pipette to enter the oil chamber. To enable the tubing from the syringe to be fitted to the micropipette while it is attached to the de Fonbrune micromanipulator, 2–3 cm at the end should be bent at 90°.

The micropipettes must be siliconized before use. To do this, fill them to about 2 cm with Repelcote, leave for a few minutes, and then rinse them out with distilled water. Filling and emptying the micropipettes is best done with a 10 ml syringe connected via the tubing and valve used in the micromanipulation itself. Again, the best way to assess the quality of the micropipettes is to test them out. Volumes of about 1 nl must be dispensed. Earlier publications advocated precise calibration of micropipettes. I have found this to be unnecessary, since in most cases equal volume additions are performed and the procedure is relatively tolerant of minor variations in buffer concentrations. I estimate these by eye, assuming the drop to be hemispherical and estimating the drop's diameter with a calibrated eyepiece graticule. A minimum of seven micropipettes will be necessary: a fresh one for each solution. Several extra micropipettes should be available, as they do break very easily.

3. Methods

3.1 Preparation of chromosomes for microdissection

The principal requirement in the preparation of chromosomes for microdissection is that the period in which they are exposed to acid fixation is kept to a minimum, around two to three minutes, in order to minimize damage to the DNA. It is important to anticipate how one is to recognize the chromosomal region of interest. In the case of polytene chromosomes, this can be achieved by familiarity with the chromosome map, as the highly detailed banding pattern can be seen under phase contrast, without staining. This is not true however in the case of metaphase chromosomes. Röhme and colleagues (11) microdissected the *t* complex from unstained murine chromosomes, and were able to recognize the appropriate chromosomal region by using a Robertsonian translocation to visualize chromosome 17 as the short arm of a metacentric chromosome in a karyotype otherwise composed of acrocentric chromosomes. Lüdecke and co-workers (2) have used GTG banding to identify human chromosomes. Kao and Yu (12) employed human–Chinese hamster ovary cell hybrids carrying human chromosomes 21 and 9 to provide a source of easily identified chromosomes 21 for microdissection.

Figure 5. Microdissection of a band from a *Drosophila simulans* polytene chromosome. The microneedle is brought close to the band to be dissected (b), and pushed through the chromosome, depositing the band to one side (c,d). By touching the tip of the needle to the microdissected fragment, it can be picked up (e) to be carried to the drop of reaction buffer.

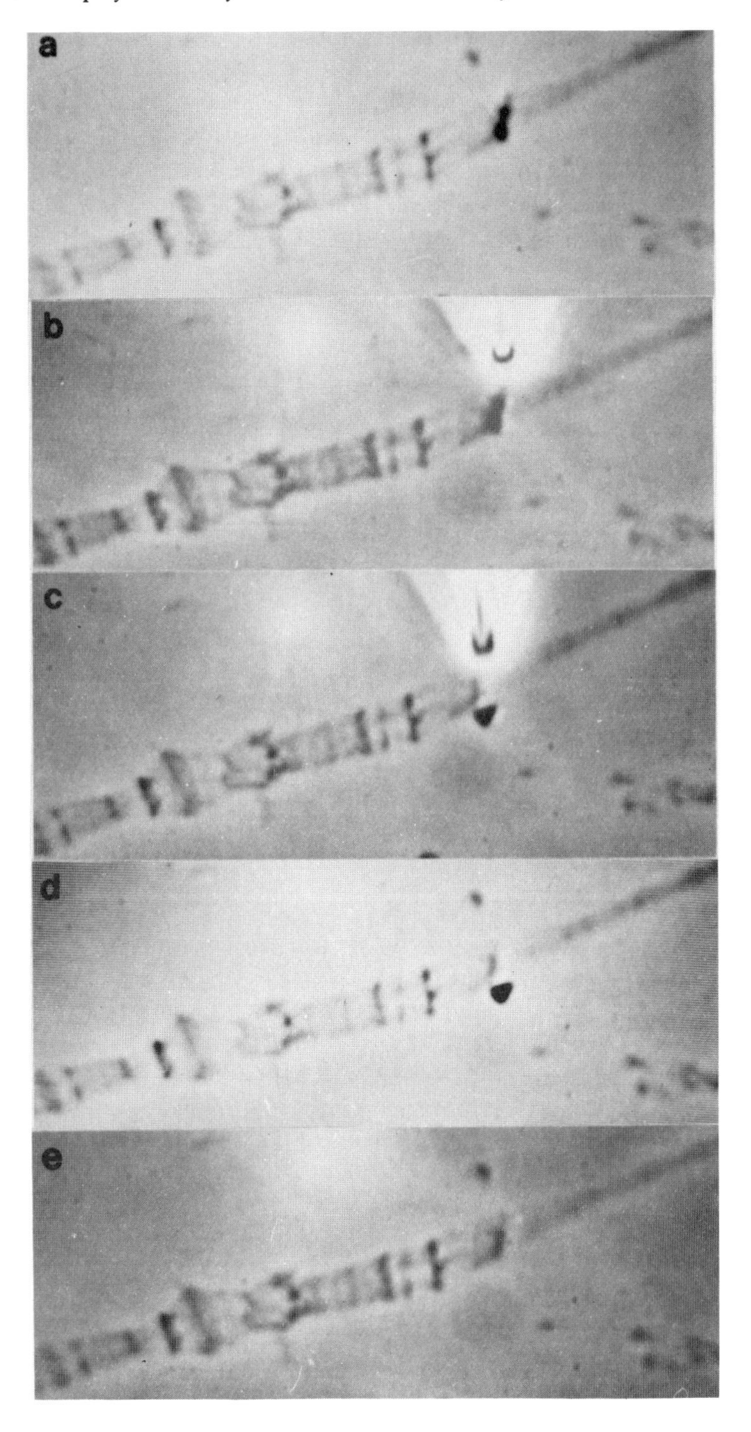

3.1.1 Polytene chromosomes

Polytene chromosomes are found in many Dipteran insects, in a variety of tissues, and the techniques for their preparation vary mostly in respect to the means for isolating the correct tissue. *Protocol 1* describes the preparation of larval salivary gland chromosomes suitable for a number of *Drosophila* species, but it has also been used in the preparation of salivary gland chromosomes of *Rhynchosciara americana* (Saunders and Santelli, unpublished) and ovarian nurse cell chromosomes of *Anopheles gambiae* (7).

Protocol 1. Preparation of salivary gland chromosomes for microdissection

Equipment and reagents

- Stereomicroscope
- Compound microscope, with dry phase × 40 objective
- Slides, siliconized 24 mm × 24 mm coverslips, and unsiliconized 22 mm × 40 mm coverslips
- Two pairs of fine forceps (Dumont No. 5; Agar Scientific Ltd, Stansted, UK, Cat. No. T5005)
- Dewar flask
- 0.7% NaCl
- 45% acetic acid
- Liquid nitrogen
- 70% ethanol
- 100% ethanol

Method

1. Dissect the salivary glands from third instar larvae, in 0.7% NaCl.
2. Transfer the glands to a drop of 45% acetic acid on a clean siliconized 24 mm × 24 mm coverslip, supported on a glass microscope slide. Leave the glands to fix for 2 min.
3. Pick up the coverslip by touching it with an unsiliconized 22 mm × 40 mm coverslip. Invert the coverslips and place them on a clean slide. A small droplet of fixative between the coverslip and the slide will prevent the coverslips from falling off the slide in step 5.
4. Spread the chromosomes by tapping in a circular movement with a pencil point. Check the chromosomes under a × 40 dry phase contrast objective. If the chromosomes are well spread proceed to step 5, if not, continue tapping the coverslip.
5. Fold blotting paper round the slide, and press heavily to squash the chromosomes. Unwrap the slide from the blotting paper, and plunge it into liquid nitrogen. When the nitrogen stops boiling, remove the slide, and flick off the siliconized coverslip with a scalpel blade. Quickly remove the large coverslip, and plunge it into 70% ethanol. After 5 min, transfer the coverslip to 100% ethanol for a further 5 min, then allow to air dry.

6. Check the chromosomes using a dry phase contrast objective. Chromosomes with good morphology that are suitable for accurate microdissection will appear flat and grey, rather than shiny and refractile.

3.1.2 Metaphase chromosomes

Metaphase chromosomes can be obtained from cultured cells, or from cytological preparations from intact tissue. Cytological squash preparations can be treated as described in *Protocol 1* as applied to polytene chromosomes, for example as with squashed preparations of *Drosophila* larval neuroblasts. In the case of *Drosophila* larval brains, these should be incubated in colchicine (10 μM in 0.7% NaCl) for 30–60 min to increase the metaphase index. Weith *et al.* (13) obtained bone marrow cells from colcemid treated mice, while methods for collection of mitotic cells from a cultured monolayer, using a micropipette have also been described (14).

Protocol 2. Preparation of metaphase chromosomes for microdissection

Reagents

- 75 mM KCl
- 3:1 methanol/acetic acid
- 70% ethanol
- PBS: 8 g NaCl, 0.2 g KCl, 1.44 g Na_2HPO_4, 0.24 g KH_2PO_4 per litre
- 80 μg/ml trypsin solution in PBS
- Giemsa's staining solution: 1:20 dilution of stock solution (Gurr's/BDH Cat. No. 35014 4M) in 10 mM phosphate buffer pH 6.8

Method

1. Incubate the cells in 75 mM KCl, to swell the cells.
2. Fix the cells in 3:1 methanol/acetic acid[a] for 5–20 min.
3. Drop the cells on to clean dry coverslips and allow the fixative to evaporate.
4. Wash and store the coverslip in 70% ethanol. If desired, the chromosomes can be GTG banded by passing the coverslips through steps 5 and 6.
5. Incubate the coverslip in trypsin solution at 37°C for 10–20 sec.
6. Rinse the coverslip in room temperature PBS, and transfer the coverslip to Giemsa staining solution for 3–5 min. Rinse in distilled water, and air dry.

[a] Kao and Yu (12) used 100% methanol to fix chromosomes with minimal damage to DNA.

3.2 Microdissection

A typical polytene chromosome band contains 1 pg of DNA. Microdissection of a single polytene band will yield sufficient DNA for a successful amplification. In the case of metaphase chromosomes, a minimum of 20 chromosome fragments are typically required, which makes the procedure more time-consuming. These fragments are accumulated in a collection drop, to which proteinase K is subsequently added to facilitate extraction of the DNA. In the case of polytene chromosome microdissection, the chromosome fragment can be transferred directly to proteinase K extraction buffer (*Protocol 3*).

Protocol 3. Microdissection of chromosome segments

Equipment and reagents

- Microcloning apparatus (see section 2.1)
- Collection buffer: 10 mM Tris–HCl pH 7.5, 10 mM NaCl
- Extraction buffer: 10 mM Tris–HCl pH 7.5, 10 mM NaCl, 0.1% SDS, 0.5 mg/ml proteinase K (the proteinase K must be added fresh)

Method

1. Using a diamond pen, cut the coverslip carrying the chromosomes to a narrow strip approximately 7.5 mm × 40 mm. Place the coverslip carrying the chromosomes face down on the oil chamber, and fill the cavity with paraffin oil. Place a siliconized coverslip, marked to indicate orientation, on the chamber next to it. Add more oil if necessary. This is the coverslip that will carry the reaction microdrops.
2. Next, place a siliconized coverslip carrying several supply drops of extraction buffer (about 0.1–0.25 μl each, dispensed using a Gilson pipette) on the oil chamber, and more oil added as required. This should be set apart from the other coverslips initially, so that the advancing oil front will be perpendicular to the long side of the new coverslip and will not sweep off the supply drops. When the oil has completely filled the space under the supply drop coverslip, push it up to the other coverslips.
3. Rinse the micropipette with extraction or collection buffer, then fill it with buffer and dispense a row of six or more 1 nl microdrops of collection or extraction buffer as required on the reaction coverslip. Estimate the volume using an eyepiece graticule. Note down the stage coordinates of each droplet in order that each reaction can be individually identified, and located. At all times during the microcloning procedure, supply drop coverslips should be positioned in front of the reaction coverslip, with the micromanipulator behind.
4. Remove the micropipette and replace it with a microneedle. Microdissection is best performed by positioning the needle against the

chromosomal region of interest and then raising it so that the tip pushes against the coverslip. It will deform and the tip will slide across the chromosome, taking a chromosome fragment with it. Lower the needle, and the chromosome fragment will probably remain on the coverslip.

5. Pick up the chromosome fragment by pressing the tip of the needle on to it. Once the fragment has attached to the needle, transfer it to the drop of extraction buffer. Do this slowly, so that the fragment does not drop off. Make sure that the needle is lower than the coverslips so that it does not touch them.
6. Bring the needle tip slowly to the droplet of extraction buffer. The fragment can be seen to either jump into the droplet, or to form viscous spools of DNA as it solubilizes. Where many fragments must be accumulated, they should be placed one by one in a microdrop of collection buffer, and an equal volume of extraction buffer added.
7. Place the oil chamber in a plastic box lined with moist filter paper. Incubate at room temperature for 30 min.

3.3 Extraction of DNA from microdissected chromosome fragments

The chromosome fragments deposited in extraction buffer are digested rapidly by the proteinase K. Obviously the proteinase K must be removed prior to the enzymatic steps that follow, as must the SDS. By phenol extracting the reaction drop, the proteinase K is inactivated. SDS is partly removed by diffusion into the aqueous saturated phenol. The phenol itself must be completely removed prior to the addition of *Sau*3AI (*Protocol 4*).

Protocol 4. Phenol extraction

Equipment and reagents

- Microdissection apparatus (see section 2.1)
- Chloroform (do not add isoamyl alcohol)
- Phenol, equilibrated with R buffer: 10 mM Tris–HCl pH 7.5, 100 mM NaCl, 10 mM $MgCl_2$, 10 mM 2-mercaptoethanol

Method

1. Remove the extraction buffer supply coverslip. Replace it with a supply coverslip[a] carrying drops of equilibrated phenol, making sure that each drop contains some aqueous phase.
2. Mount a micropipette on the micromanipulator. Rinse the micropipette with phenol, then fill it with phenol. Dispense approximately 4–8 μl round the droplet of extraction buffer. Incubate at room temperature for 3–5 min. Remove the phenol phase using the micropipette.

Protocol 4. *Continued*

3. Replace the micropipette, and repeat step 2 twice more.
4. Remove as much phenol as possible after the final phenol extraction.
5. Fill a micropipette with chloroform,[b] then mount it on the micromanipulator.
6. Remove residual traces of phenol by squirting a jet of chloroform past each droplet. Great care must be taken not to dislodge the reaction droplet from the coverslip, or to touch the droplet with the pipette tip.
7. The oil in the chamber is now contaminated with phenol and chloroform, so the coverslip bearing the reaction droplets must be transferred to a fresh oil chamber. Place an unsiliconized coverslip across a second oil chamber, and fill the chamber with oil. Bring the two chambers together, and slide the reaction coverslip across to the new oil chamber. Ensure the coverslip is flanked by unsiliconized coverslips.

[a] Use unsiliconized coverslips, as phenol will not adhere sufficiently to siliconized coverslips.
[b] Supply drops of chloroform are impossible, as chloroform is miscible with oil.

3.4 Restriction enzyme digestion

To reduce the microdissected DNA fragments to a manipulable size, it is digested with a restriction enzyme. The original microcloning protocols used *Eco*RI to generate fragments for cloning into λ vectors. However many *Eco*RI fragments are too large for reliable PCR amplification, and more frequent cutting restriction enzymes are therefore used. The enzyme of choice in many cases is *Sau*3AI, or its isoschizomer *Mbo*I. These enzymes are available in high concentration from a number of suppliers, and are heat inactivatable, removing the necessity of a further phenol extraction at this stage (*Protocol 5*). In some cases it is worth using an alternative restriction enzyme, for example *Sau*3AI is inadvisable if a high copy number repetitive element in the genome under study contains one or more *Sau*3AI sites.

Protocol 5. Digestion with *Sau*3AI

Equipment and reagents

- Microdissection apparatus (see section 2.1)
- 4 × R buffer: 40 mM Tris–HCl pH 7.5, 0.4 M NaCl, 40 mM $MgCl_2$, 40 mM 2-mercaptoethanol
- *Sau*3AI, at a concentration of at least 50 U/μl (Amersham International)

Method

1. Prepare 2 μl of a working dilution of *Sau*3AI by mixing 1 μl concentrated enzyme with 1 μl of 4 × R buffer. The resulting enzyme concentration

should be 25 U/μl. Dispense supply drops on a siliconized coverslip and place the coverslip on the oil chamber adjacent to the reaction coverslip.

2. Mount a fresh micropipette on the micromanipulator. Rinse the pipette in one of the supply drops, and then fill with enzyme solution. Locate the reaction drops, and dispense a droplet of equal volume adjacent to each of them. The two drops should fuse if they are touching, and will mix by diffusion.
3. Place the oil chamber in the humid box, and incubate for 1.5–2 h at 37°C.
4. To inactivate the *Sau*3AI, incubate the humid chamber at 70°C for 20 min.

3.5 Ligation of oligonucleotide adapters

The oligonucleotide adapters are double-stranded, with a cohesive terminus compatible with the restriction enzyme chosen for *Protocol 5*. The published method of Saunders *et al.* (3) used phosphorylated adapters to permit double-stranded ligation to the chromosomal DNA fragments (*Protocol 6*). In order that adapter dimers are not present in the amplification step, they must be cleaved following the ligation. The adapter shown in *Figure 1* creates a *Bcl*I site when self-ligated enabling adapter dimers to be eliminated by digestion with *Bcl*I. However, this procedure cleaves 25% of the genomic DNA:adapter junctions, so 50% of the genomic fragments will not be amplified. It is simpler to use unphosphorylated adapters. These will only allow single-strand ligations to the *Sau*3AI fragments but selection of the appropriate thermal cycle program can overcome this problem (see section 3.6), by permitting nick translation to occur. This makes the *Bcl*I digestion step unnecessary, so representation is increased.

Protocol 6. Adapter ligation

Equipment and reagents

- Microdissection apparatus (see section 2.1)
- T4 DNA ligase, 1 U/μl
- Adapter solution: 10 μM adapters in 10 mM Tris–HCl pH 7.5, 10 mM $MgCl_2$, 5 mM 2-mercaptoethanol, 2 mM ATP

Method

1. Following the heat inactivation of *Sau*3AI in *Protocol 5*, step 5, remove the coverslip bearing supply drops of *Sau*3AI. Replace it with a siliconized coverslip carrying three supply drops of adapter solution, and three drops of ligase.

Protocol 6. *Continued*

2. Pipette an equal volume of adapter solution adjacent to each reaction drop. Allow the drops to fuse.
3. Introduce ligase to each reaction drop with the micropipette. To do this, the tip of the pipette must touch the reaction drop. Ideally therefore, a fresh pipette should be used for each reaction, to avoid cross-contamination between reactions. The ligase should be visible as it enters the reaction drop. About 50% drop volume should be added.
4. Incubate the oil chamber overnight at 4°C in a humid box.

3.6 PCR amplification

The polymerase chain reaction is extremely sensitive to contamination. Care should be taken at all times to use clean and sterile materials and solutions. Wear gloves at all times, and take precautions to avoid carry over of DNA or solutions, particularly avoiding cross-contamination with samples previously amplified using this procedure (*Protocol 7*).

Protocol 7. PCR amplification of microdissected DNA

Equipment and reagents

- Thermal cycler
- 10 × reaction buffer: 100 mM Tris–HCl pH 8.3, 500 mM KCl, 15 mM $MgCl_2$
- 0.1% gelatin
- 100 μM primer
- 10 mM dNTPs (10 mM each nucleotide triphosphate)
- *Taq* polymerase (5 U/μl)
- Paraffin oil

Method

1. Place the oil chamber in a 90 mm diameter Petri dish, and fill the dish with paraffin oil. Take care not to sweep the reaction drops off with the flow of oil. The reaction drops should be visible using a dissecting microscope.
2. Slip one of the flanking coverslips off the oil chamber, then move the reaction coverslip to the edge of the chamber. Flip the coverslip over and place on the bottom of the Petri dish, drops uppermost. This is quite an easy manipulation, but does require practice to avoid losing the reaction drops.
3. Place 10 μl sterile distilled water in microcentrifuge tubes. Pick up 1–2 μl with a pipette. The reaction drop can be picked up by expelling a little water on to the drop and removing it again. Place the sample in the tube. Note that the ligation can become quite viscous, try not to leave DNA adhering to the coverslip.

4. Set-up the PCR reaction by adding the following:
   - 10 × reaction buffer 10 μl
   - 10 mM dNTPs 2 μl
   - 100 μM primer 1 μl
   - sterile water 76.5 μl
   - *Taq* polymerase (5 U/ml) 0.5 μl
   - total volume 90 μl

   Cover the reactions with a layer of paraffin oil before thermal cycling.
5. Typically a suitable program is:
   - 37°C 5 min
   - 55°C 5 min
   - 72°C 5 min

   Then 30 cycles of the following three steps:
   - 94°C 1 min
   - 55°C 1 min
   - 72°C 3 min
6. After the program is complete, add 100 μl of chloroform to the tube, which inverts the phases, and remove the aqueous phase to a fresh tube.

3.7 Characterization of amplified material

The DNA amplified from microdissected chromosome fragments represents a substantial sequence complexity. Consequently, gel electrophoresis reveals a smear rather than a discrete band. In the case of *Sau*3AI digested chromosomal DNA, the resulting smear is of mean fragment length 300 bp. In some cases, the presence of repetitive DNA can result in bands superimposed upon a smear on the gel. However, if only a few, clean, bands are seen, it is likely that the experiment has failed. Should the DNA look as expected, the easiest means by which the quality of the material can be assessed is by *in situ* hybridization to chromosomes, especially in the case of polytene chromosomes.

3.7.1 Synthesis of probes for *in situ* hybridization

Probes can be made from DNA amplified by the polymerase chain reaction (PCR), using additional PCR cycles in the presence of biotinylated nucleotide (*Protocol 8*). As an alternative biotin labelled probes can be synthesized by random priming (15).

Protocol 8. Synthesis of probes from PCR amplified DNA

Reagents

- PCR buffer and primer (see *Protocol 7*)
- dNTP solution: 10 mM each of dCTP, dGTP, dATP
- Biotin-16-dUTP (Boehringer)
- 2 × hybridization buffer: 8 × SSC, 2 × Denhardt's solution, 20% dextran
- sulfate, 0.4% denatured salmon sperm DNA

Method

1. Remove unincorporated nucleotides from the amplified DNA, for example by gel electrophoresis.
2. Set-up the PCR reaction, using the same reaction conditions as used to amplify the DNA (*Protocol 7*)

• 10 × reaction buffer	5 μl
• 10 mM dNTPs (A, G, C)	0.5 μl
• 1 mM biotin-16-dUTP	1 μl
• 100 μM primer	0.5 μl
• PCR amplified DNA	2 μl
• sterile water	40.75 μl
• *Taq* polymerase (5 U/ml)	0.25 μl
• total volume	50 μl

   Five to ten cycles of synthesis are usually sufficient. Increasing the length of the polymerization step from 3 min (as in *Protocol 7*) to 10 min is advised, since the concentration of biotin-16-dUTP is low.
3. Ethanol precipitate the labelled DNA and resuspend it in 50 μl sterile distilled water to which 50 μl 2 × hybridization solution is then added.

3.7.2 *In situ* hybridization to polytene chromosomes

The power of *in situ* hybridization to polytene chromosomes is due largely to the size of these chromosomes and consequently the high degree of resolution they offer in cytogenetic mapping. The use of radiolabelled probes has now been largely superseded by non-radioactive signal detection systems, generally using biotin substituted probes which offer greater resolution, since there is less scatter of signal with immunochemical and immunofluorescent detection than with silver grains. *Protocol 9* is devised for *Drosophila* larval salivary gland polytene chromosomes, but should be suitable for polytene chromosomes from other species and tissues, although minor variations may be required in the preparation of chromosomes. Similar minor variations can be found in *Protocol 10* for *in situ* hybridization to metaphase chromosomes.

Protocol 9. *In situ* hybridization to polytene chromosomes

Equipment and reagents

- 45% acetic acid
- 1:2:3: fix: one part lactic acid: two parts water: three parts acetic acid
- Slides and coverslips
- Liquid nitrogen

Method

1. Dissect out tissue in a drop of 0.7% NaCl, and transfer them to a drop of 45% acetic acid. Allow to fix for approximately 30 sec.
2. Transfer the glands to a drop of 1:2:3 fixative on a clean siliconized coverslip. Fix for 3 min then pick up the coverslip with a clean slide, by touching it to the drop. The slide does not have to be coated or 'subbed' before use.
3. Spread the chromosomes by tapping the coverslip with a pencil in a circular motion. Check the chromosomes using phase contrast microscopy. When the chromosomes are suitably spread, fold the slide in blotting paper, and press gently to remove excess fix. Leave the slide at room temperature for at least 1 h or overnight. This step squashes the chromosomes as the fix evaporates, and the coverslip sinks towards the surface of the slide. Alternatively, the slide can be squashed firmly between blotting paper, and frozen immediately. Take care not to allow the coverslip to slide sideways, or the chromosomes will be overstretched.
4. Freeze the slide in liquid nitrogen. While the slide is still frozen, flip off the coverslip with a scalpel blade and proceed to step 5.
5. Place the slide in 70% ethanol for 5 min. Transfer the preparation through two 5 min changes of 96% ethanol, and air dry. The chromosomes can be stored desiccated at room temperature. Use only slides with good quality chromosomes, those which appear flat and grey, with clear banding. Chromosomes which appear bright and reflective under dry phase examination will have poor morphology, hindering accurate interpretation. The region of the slide where the chromosomes are located should be visible when the slide is dry, and should be marked with a diamond pencil on the reverse side of the slide.
6. Incubate the slides in 2 × SSC at 65°C for 30 min. This step is intended to help preserve the morphology of the chromosomes.
7. Transfer the slides to 2 × SSC at room temperature for 10 min, then denature the chromosomes by incubating the slides in 70 mM NaOH for 2 min. The 70 mM NaOH must be freshly made.
8. Rinse the slides in 2 × SSC.

Protocol 9. *Continued*

9. Dehydrate through ethanol as described in *Protocol 1*, step 5 above, and air dry. The slides should be used the same day.
10. Boil the probe for 3 min, and quench on ice. Check the volume after boiling, and restore to the initial volume with sterile distilled water. Pipette 20 μl of the probe (in 1 × hybridization buffer, *Protocol 8*) on to the chromosomes. Cover the chromosomes and probe with a clean siliconized coverslip. There is no need to seal the coverslip.
11. Place the slides in a plastic box lined with moist tissue to prevent evaporation from the preparation, seal the lid, and place in a 58°C incubator overnight.
12. Remove the slides from the humid box. Dip them in 2 × SSC to allow the coverslip to slide off, then wash them in 2 × SSC, 53°C, for 1 h. Three changes of wash solution should be made.

Protocol 10. *In situ* hybridization to metaphase chromosomes

Reagents

- 70% formamide, 2 × SSC
- 70% ethanol
- 100% ethanol
- 2 × SSC

Method

1. Prepare chromosomes by a method appropriate to the material. A similar protocol to that used in preparation of chromosomes for microdissection can be used.
2. Bake chromosomes at 65°C for 30 min to harden the chromosomes. Denature the chromosomes by incubation in 70% formamide, 2 × SSC at 70°C for 2 min. Remove slides to 70% ethanol for 5 min. Pass through two 5 min washes in 100% ethanol.
3. Proceed as in steps 10–12 of *Protocol 9*.

A variety of methods are available by which *in situ* hybridization signals can be visualized. Avidin– or streptavidin–HRP conjugates used with the HRP substrate diaminobenzidine (DAB) produce a clear signal on polytene chromosomes. This signal is dark brown, and insoluble, and so slides prepared in this way are durable and can be stored. A more sensitive approach is to use a fluorescent tagged avidin or streptavidin. While there is a significant gain in sensitivity, the preparations do not store well. In general, HRP–DAB preparations are sensitive enough for polytene chromosomes and are therefore preferable. The fluorescent detection system is advisable for hybridization to mitotic chromosomes due to the increased sensitivity.

Protocol 11. Signal detection with avidin–HRP conjugate

Reagents

- PBS: 8 g NaCl, 0.2 g KCl, 1.44 g Na_2HPO_4, 0.24 g KH_2PO_4 per litre
- PBS–TX: PBS containing 0.1% Triton X-100
- Extravidin–horse-radish peroxidase conjugate (Sigma)
- Extravidin–HRP dilution buffer: PBS containing 10 mg/ml BSA, 5 mM EDTA
- 0.5 mg/ml diaminobenzidine (Sigma, Cat. No. D 9015), 0.01% hydrogen peroxide, in PBS – the hydrogen peroxide should be added immediately prior to use
- Giemsa's stain: 1:20 dilution of Giemsa's solution (Gurr's-BDH) in 10 mM phosphate buffer pH 6.8

Method

1. The slides should be taken from the final wash in *Protocol 9* and washed twiced for 5 min in PBS. Wash the slides for 2 min in PBS–TX. Rinse in PBS. Do not allow the slides to dry out during signal detection.
2. Make a 1:250 dilution of extravidin–horse-radish peroxidase conjugate in dilution buffer. Apply 50 µl to the chromosomes and cover with a 22 × 50 mm coverslip. Replace the slides in the humid box, and incubate at 37°C for 30 min.
3. Wash off the unbound conjugate by passing the slides through the PBS and PBS–TX washes described in step 1. Drain the slides, but do not allow them to dry.
4. Place 50 µl DAB solution on to the chromosomes, and cover with a 22 × 50 mm coverslip. This solution should be made up fresh since hydrogen peroxide decays rapidly. Take care when working with diaminobenzidine, as it is a potent carcinogen. Incubate at room temperature for 10–15 min in the humid box. Rinse the slides in PBS and examine under phase contrast. The signal appears blackish-brown, sometimes quite refractile in strong cases. If the signal seems weak, add more DAB solution, and incubate for longer. If the signal is strong enough, rinse the slide well with distilled water.
5. Stain in Giemsa's stain for 1 min, rinsing off excess stain in running water for a few seconds. Allow the slides to air dry. Check that the staining is sufficiently intense. Overstained chromosomes can be destained in 10 mM sodium phosphate buffer pH 6.8, and understained chromosomes can be re-stained. The preparation should be mounted under a siliconized coverslip with DPX mounting medium. DPX is a xylene soluble mountant, which does not affect either the Giemsa stain or DAB deposit, and the slides should last for many years. It is convenient to seal the edges of the coverslip with nail varnish to prevent immersion oil from seeping under the coverslip. The preparations should be examined under phase contrast. Photographic reproduction is best with colour film.

Protocol 12. Signal detection using avidin–fluorescein

Reagents

- PBS, PBS–TX
- Avidin–fluorescein conjugate (Jackson)

Method

1. Wash the slides in PBS as described in *Protocol 9*, step 12. Pass the slides through two 5 min washes in PBS and one 2 min wash in PBS–TX. Rinse the slides in PBS. Do not allow the slides to dry out.
2. Make a 1:50 dilution of avidin–fluorescein in PBS. Pipette 50 μl on to the chromosomes, cover with a 22 × 50 mm coverslip, and incubate in a dark humid box at room temperature for 30 min.
3. Pass the slide through the PBS and PBS–TX washes as step 1 above, drain the slide, and mount in mounting medium, under a siliconized coverslip. The mounting medium contains propidium iodide to stain the chromosomes.
4. Seal the coverslip with nail varnish. These preparations do not last as long as those described in *Protocol 11*, but will keep for several weeks at 4°C in the dark, if sealed as described. They are best examined using a microscope with facility for image merging. A confocal microscope is ideal.

3.7.3 Cloning amplification products

The adapter shown in *Figure 1* contains restriction sites for *Hin*dIII and *Eco*RI, in addition to *Sau*3AI. It is therefore possible to clone the amplified DNA fragments into a wide range of plasmid or bacteriophage vectors to produce a chromosome region-specific minilibrary of cloned *Sau*3AI fragments. Relevant cloning techniques can be found in *DNA cloning: a practical approach: core techniques*. It is possible to amplify DNA fragments that are not derived from the microdissected chromosomes. Such contaminating DNA can be traced back to bacterial contamination of solutions or chromosome coverslips, or some other contamination. The source of amplified DNA can be checked by Southern blot hybridization (15) to genomic DNA. In addition, this analysis will determine whether a particular fragment is single or multicopy, and the number of copies of a multicopy fragment present in the genome can be estimated.

4. Applications

4.1 Cloning specific loci

Chromosome microdissection can be a valuable component of strategies for molecular cloning. The simplest approach is to use microdissection to provide a region-specific probe with which to select genomic clones corresponding to the known cytological position of the gene of interest. In principle, all the clones required to cover the region can be identified on hybridization to a genomic library, though this is obviously subject to the completeness of the representation of the region in the library, and the accuracy of the microdissection. All the clones produced in this way must be checked by *in situ* hybridization to verify their site of origin: some clones will be selected solely because they contain a repetitive element present in the microdissected probe, and others may contain chimeric inserts. These clones can be used to initiate a chromosome walk. Probes prepared by microdissection have been used in this way in *Drosophila* genome mapping.

If the time and tissue of expression of a gene are known in addition to its chromosomal location, a probe prepared by microdissection may be used to screen an appropriate cDNA library. Conversely, the microdissected DNA might be cloned, and screened with an appropriate cDNA probe. This strategy will work best with the high-resolution afforded by polytene chromosomes.

Transposon tagging has been widely used to clone genes, particularly in *Drosophila*. Mutants produced by hybrid dysgenesis should in principle be due to the insertion of the relevant transposable element either within the gene or in close proximity to it. Often, many elements have been mobilized, and these are removed by recombination prior to constructing a genomic library from the mutant stock which is then screened with the transposon probe. However, a library constructed using the original mutagenized stock can be screened with both transposon and a microdissected probe. Clones identified by both probes correspond to those containing an element inserted within the region of interest (16).

4.2 Microsatellites, RFLPs, and STSs

Microsatellites, RFLPs, and STSs can be isolated from minilibraries. STSs are essentially short single copy fragments of known chromosomal location which can be sequenced to yield a marker for aligning genomic YAC, cosmid, or P1 clones. STSs and clones identifying restriction fragment length polymorphisms (RFLPs) must be identified by analysis of individual clones selected from minilibraries. Microsatellite clones can be selected from minilibraries, derived as described in section 3.7.3, by probing with appropriate simple sequence repeats (typically dinucleotide repeats).

Clones containing microsatellites or useful STSs or RFLPs are essentially 'pre-sorted' by their chromosomal site of origin, and isolating clones in this way can greatly facilitate the production of a collection of markers evenly distributed through the genome. Of course the distinction between these three classes of clone is blurred: a microsatellite marker can also be used as an STS or to reveal an RFLP by virtue of its single copy flanking sequences.

References

1. Scalenghe, F., Turco, E., Edström, J. E., Pirrotta, V., and Melli, M. (1981). *Chromosoma*, **82**, 205.
2. Lüdecke, H.-J., Senger, G., Claussen, U., and Horsthemke, B. (1989). *Nature*, **338**, 348.
3. Saunders, R. D. C., Glover, D. M., Ashburner, M., Sidén-Kiamos, I., Louis, C., Monastirioti, M. *et al.* (1989). *Nucleic Acids Res.*, **17**, 9027.
4. Johnson, D. H. (1990). *Genomics*, **6**, 243.
5. Sidén-Kiamos, I., Saunders, R. D. C., Spanos, L., Majerus, T., Trenear, J., Savakis, C., *et al.* (1990). *Nucleic Acids Res.*, **18**, 6261.
6. Kafatos, F. C., Louis, C., Savakis, C., Glover, D. M., Ashburner, M., Link, A. J., *et al.* (1991). *Trends Genet.*, **7**, 155.
7. Zheng, L., Saunders, R. D. C., Fortini, D., della Torre, A., Coluzzi, M., Glover, D. M., *et al.* (1991). *Proc. Natl Acad. Sci. USA*, **88**, 11187.
8. Ponelies, N., Bautz, E. K. F., Monajembashi, S., Wolfrum, J., and Greulich, K. O. (1989). *Chromosoma*, **98**, 351.
9. Frey, M., Koller, T., and Lezzi, M. (1982). *Chromosoma*, **84**, 493.
10. Wesley, C. S., Ben, M., Kreitman, M., Hagag, N., and Eanes, W. F. (1990). *Nucleic Acids Res.*, **18**, 599.
11. Röhme, D., Fox, H., Herrmann, B., Frischauf, A.-M., Edström, J.-E., Mains, P., *et al.* (1984). *Cell*, **36**, 783.
12. Kao, F.-T. and Yu, J.-W. (1991). *Proc. Natl Acad. Sci. USA*, **88**, 1844. Saunders (1990). *Bioessays*, **12**, 245.
13. Weith, A., Winking, H., Brackmann, B., Boldyreff, B., and Traut, W. (1987). *EMBO J.*, **6**, 1295.
14. Claussen, U. (1980). *Hum. Genet.*, **54**, 277.
15. Alphey, L. M. and Parry, H. (In press) In *DNA cloning: a practical approach* Vol. 1, *Core techniques*, (ed. D. M. Glover and B. D. Hames), pp. 121–41. IRL Press, Oxford.
16. Philp, A. V., Axton, J. M., Saunders, R. D. C., and Glover, D. M. (1993). *J. Cell Sci.*, **106**, 87.

6

Databases, computer networks, and molecular biology

R. FUCHS and G. N. CAMERON

1. Introduction

Since M. Dayhoff established the first public macromolecular sequence data collection in the mid-1960s databases have become indispensable research tools in molecular biology (1). The number of databases of interest to biologists has grown steadily since then, and in 1993 the 'database of databases' LiMB (Listing of Molecular Biological Databases) presented more than a 100 databases in this area (2). Regular consultation of databases for nucleotide or protein sequence information, mapping data, or the description of the functional role of a particular protein in the living cell is now essential to the work of any molecular biologist.

All biological databases today are available, often exclusively, in electronic form. Thus a certain amount of computer literacy is crucial for a modern biologist, even though computing has so far not been a key element of biological training and education. A biologist who mastered the principles of using a computer to support him in his paperwork may still face apparently insurmountable obstacles when he tries to find a small piece of relevant information in the mass of information provided by current databases.

Accessing databases and extracting interesting information efficiently is getting increasingly important but also more difficult, due to the rapid increase in the size and number of databases and their growing diversity. Databases of several hundred megabytes of data are not uncommon, and progress in genome sequencing and mapping will soon result in databases of several gigabytes in size (3). At the same time, large general databases such as the nucleotide sequence databases and gene mapping databases are being complemented by a growing number of smaller specialized data collections with information on, for instance, protein properties, diseases, or regulatory elements. Unfortunately, the development of these databases is not always well co-ordinated, and the user can be confronted with, and confused by, a wide variety of different data models, database management systems, formats, and distribution media.

In this chapter we describe recent developments in data access and database retrieval technology. Our goal is to provide biologists with some help in manipulating biological databases and finding the information they need. A thorough description of individual biological databases is out of scope; the reader is referred to the literature (2, 4–6). Instead we concentrate on general issues of database maintenance and describe access methods which are generally applicable. The problems associated with the identification of similar or related sequences in a sequence database are also covered elsewhere in detail (7) and will not be explored here.

Our focus is on scientists working in laboratories with modest informatics investment, typically equipped with small computer systems such as MS–DOS or Apple Macintosh computers. In contrast to scientists at larger centres these researchers get less assistance from dedicated bioinformatics staff responsible for database installation, maintenance, and user support.

We start with a discussion of the problems associated with the local maintenance of data collections and present some solutions based on CD–ROM (compact disk read-only memory) technology. The emphasis of this chapter, however, is on computer network access to biological databases. Today we can assume that almost any university or company can provide a local researcher with access to world-wide computer networks. We will demonstrate that new tools offer biologists easy and convenient access to a wealth of information stored in databases around the globe, and that these network approaches, together with CD–ROM-based local database maintenance, can make working with biological databases a lot simpler.

2. Local database management

When a researcher needs to query a public domain database for information, the most obvious approach is to get a copy of the database and install it on a local computer system. However, it will quickly become clear that this strategy is not without its problems.

Essential databases such as the nucleotide sequence database have reached a size that makes it difficult to store them on computer systems widely used in molecular biology such as MS–DOS or Apple Macintosh computers. The current (December 1993) release of the EMBL nucleotide sequence database at 450 megabytes can already pose storage problems, and it doubles in less than two years. Even for smaller databases it is not unusual that they occupy several megabytes of disk space. Considering that most scientists need access to several databases, storage may soon become a problem for small computer systems requiring a move to more powerful computer platforms such as UNIX or VMS systems, associated with a whole new set of complications, particularly in system management.

Databases are not static. They change rapidly and most biological databases are updated regularly. Some databases, such as the nucleotide sequence database, are updated on a daily basis. As a consequence, significant efforts must be invested if local copies of databases are to be kept current. Maintenance of a local database collection can easily become a chore, especially if the databases are obtained from different sources and their revision cycles are not synchronized.

Even complete and up to date databases are useless without the tools to access them—software for querying and retrieving information. Many excellent programs, both commercial and public domain, are now available which provide such capabilities. Often these systems come with several databases pre-installed. This turns out to be a mixed blessing—most software vendors have to process the databases as supplied by their producers in order to make them compatible with the software tools, resulting in delays of weeks or months. The installed databases can therefore be long out of date even by the time they are delivered.

The information needs of today's molecular biologist typically require access not only to several databases but also to several software packages. This means that every new release of every database needs to be installed for all the software packages (as only few use the databases in their native format), resulting in extensive disk space requirements.

Interesting approaches that tackle some of the problems of local database maintenance have been made possible by the development of a new storage medium for databases, CD–ROM, and by recent developments in computer network access methods. In the following we describe some tools and methods that can assist scientists, and owners of small computer systems in particular, to efficiently manage and use databases on their local computers.

2.1 Databases on CD–ROM

Only a few years ago CD–ROM was introduced into molecular biology as a data distribution medium and since then has established itself rapidly as a convenient way of working with large databases. Database producers such as the Data Library at the European Bioinformatics Institute (EBI), the National Center for Biotechnology Information (NCBI), or the Protein Identification Resource (PIR-International) now use CD–ROM as their main means of data distribution. With a storage capacity of about 600 megabytes it is possible to fit the whole nucleotide sequence database on one disk. CD–ROM production is cheap and therefore the cost of CD–ROM database products is normally very low. Another advantage of CD–ROM is the existence of an international standard (ISO 9660) that allows the same disk to be read on most hardware platforms so that it is possible for a database producer to support DOS, UNIX, VMS, and Macintosh users with only one version of a CD–ROM. *Table 1* provides a list of major suppliers of molecular biological databases on CD–ROM. In addition to the publicly funded organizations listed, which distribute their CD–ROMs for nominal fees, most

Table 1. Molecular biological databases on CD–ROM

Product	Contact address
EMBL nucleotide sequences, SWISS–PROT protein sequences, other databases.	EBI, Hinxton, Cambridge CB10 1RQ, UK. Phone +44–1223–494400
GenBank nucleotide sequences, protein sequences, EST database, MEDLINE molecular sequence data subset, other databases.	Natl. Center for Biotechnology Information, NLM, 8600 Rockville Pike, Bethesda, MD 20894, USA. Phone +1–301–496–2475.
PIR protein sequences, GenBank nucleotide sequences, other databases.	Natl. Biomed. Research Found., 3900 Reservoir Rd. NW, Washington DC 20007, USA. Phone +1–202–687–2121.
PIR protein sequences, GenBank nucleotide sequences, other databases.	Martinsried Institut für Protein-sequenzen, MPI für Biochemie, 8033 Martinsried, Germany. Phone +49–89–8578–2657.
PIR protein sequences, GenBank nucleotide sequences, other databases.	Japan Intl. Protein Information Database, Science University of Tokyo, 2669 Yamazaki, Noda 278, Japan. Phone +81–48–124–1501.
PDB protein structure data.	Protein Data Bank, Chemistry Department, Building 555, Brookhaven National Laboratory, Upton, NY 11973, USA. Phone +1–516–282–3629.

producers of commercial molecular biological software now supply databases on CD–ROM, too.

The low cost of CD–ROM renders this medium particularly interesting for 'low-end' users who previously had little opportunity to use large databases but can now cheaply buy CD–ROM drives for their laboratory PCs or workstations. The usefulness of CD–ROM is further increased by the availability of software products allowing database queries and information retrieval directly on the CD–ROM. Most database producers supply advanced query and retrieval software on their CD–ROM or as a separate product, and we will describe some of these programs below.

Although the speed of access to a CD–ROM is inevitably slower than access to a magnetic disk, the large storage capacity of a CD–ROM encourages extensive indexing by database suppliers, which can, for many applications, overcome the limitations of inherent slow access.

A disadvantage of CD–ROM is that disk mastering and pressing is typically contracted out, with production scheduling often imposing delays. This makes it more suitable for periodic releases of databases every few months than for rapid updates. For many researchers this time lag is not acceptable and more immediate access to latest data is required (see section 3). If access to latest data is not crucial, however, or can be achieved by other means—such as using on-line services (see section 3.2)—a laboratory can be served well with regular updates of its data collections on CD–ROM.

2.2 CD–ROM-based database query software

With the switch from magnetic tape to CD–ROM as the main data distribution medium, most database producers have also started to utilize the capabilities of the new medium by developing application software that allows the querying of the databases on CD–ROM and the extraction of interesting information. These programs are usually bundled with, and optimized for, use with the databases provided on the same CD–ROM. Programs such as Entrez (NCBI), EMBL-Search (EBI), CD-SEQ (EBI), or Atlas (PIR) all provide basic database query functions and enable researchers to look up information by name, keywords, taxonomy, free text, etc., and to retrieve relevant entries from the databases.

Several databases can fit easily on one CD–ROM disk. This invites the development of software that enables searches of multiple databases simultaneously or that utilizes links between the databases, and this area is what distinguishes most of the currently available retrieval systems. For example, because their CD–ROM combines sequence data with abstracts from the MEDLINE literature database NCBI's Entrez software provides researchers with an elegant mechanism to move easily from an entry in the nucleotide sequence database to a MEDLINE literature database entry that describes a paper cited in the sequence database entry. With PIR's Atlas program one can query several sequence databases simultaneously, while the EMBL-Search software utilizes cross-references between databases such as EMBL, SWISS-PROT, EPD, or PROSITE and thus enables a researcher to quickly look up related information in several databases.

As an example, *Figure 1* shows a typical query using the EMBL-Search software running on an Apple Macintosh which finds all entries in the EMBL nucleotide sequence database release 36 containing data on ribosomal RNA from cyanobacteria. Within three seconds the software identified eight matching entries. The entries found can be displayed on screen, printed, or copied to disk files. By simple double-clicking on a line that contains a cross-reference to the SWISS-PROT database, the user can pop up a new window with the protein sequence that corresponds to a given nucleotide sequence. From there he may then move on to the PROSITE database for a detailed description of the protein family. This way, related information from other databases can be quickly identified which otherwise may have been overlooked.

2.3 CD–ROMs on local area networks

The usefulness of CD–ROM can be further increased by making them available on a local computer network. Thanks to significant advances in network

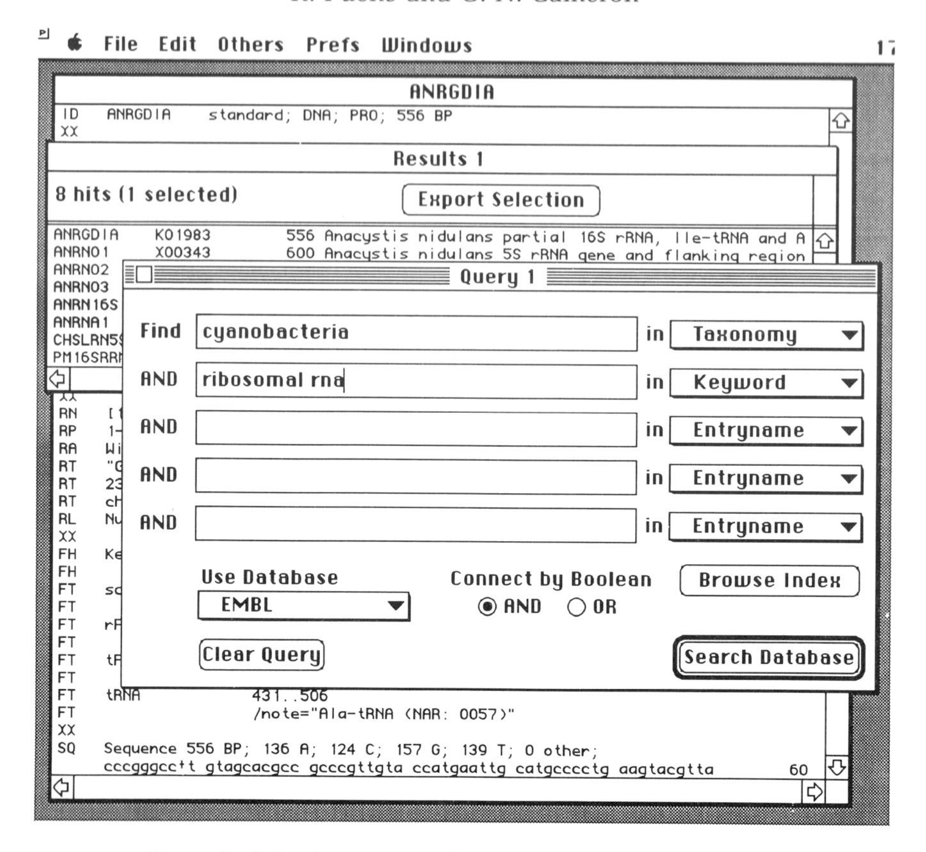

Figure 1. A database query using EBI's EMBL-Search software.

technology it is now easy to connect several local computer systems and let them share central resources. Using cheap wiring solutions and software such as AppleShare or Novell it is possible to make very efficient use of CD–ROM products by mounting one or more CD–ROMs on a central 'server' system so that scientists can access them from their personal workbench or desktop workstations. This approach is obviously more cost-effective than equipping each individual computer system with CD–ROM drives and buying multiple copies of a CD–ROM product, which is often not feasible for small laboratories. The performance of such a network configuration is dependent on network quality and number of simultaneous users.

A description of the installation at the European Molecular Biology Laboratory in Heidelberg may serve as an example for the usefulness of a CD–ROM network. Several hundred Macintosh computers are connected via AppleTalk or Ethernet cabling to a central Macintosh server. Attached to this server computer are five CD–ROM drives which hold releases of the EMBL sequence database, the MEDLINE literature database, and Current Contents, while the necessary retrieval software for these information products is stored at and run from the local computers. This service has been very successful and scientists at EMBL make extensive use of it. With a single subscription to a database it is thus possible to support a large number of users efficiently.

3. Network information resources

While CD–ROM products constitute invaluable information resources, their inherent limitations—in particular storage restrictions, access constraints, and prolonged update cycles—call for more advanced methods for data and information access. With dramatic improvements in computer network technology, scientists can now make effective use of high-speed networks and network information resources to supplement or replace local databases. International computer networks provide access to central computing centres, remote database servers, and sequence analysis facilities. Most interestingly, the flexibility of electronic communication allows the retrieval and use of genomic information which is far more up to date than that delivered on physical media such as magnetic tape or CD–ROM.

3.1 Network access

Throughout this section it is assumed that the reader's computer system is connected to an international computer network. Space limitations do not allow us to cover technical aspects of establishing network connectivity since details depend too much on local peculiarities. Readers are advised to contact the local computer centre associated with their campus, institute, or organization if they do not have a network connection yet.

The rest of the text assumes that the computer user is connected to the world's largest computer network, the Internet, which links more than 1 000 000 computers. The Internet provides four basic network functions, based on the TCP/IP protocol suite:

(a) *Terminal access*; the ability to connect (login) to a remote host system and make one's own computer act as a simple input/output device while executing programs on the other machine.
(b) *Electronic mail*; the ability to exchange messages between users on different machines.
(c) *File transfer*; the ability to exchange arbitrary files between two computer systems.
(d) *Client/server communication*; the ability to let two programs running on different computer systems communicate with each other.

The following sections show examples of network information services that utilize one or the other of these options to make databases and other information available to biologists via computer networks. Good examples of using network information resources and tools to solve biological problems are given in ref. 8.

3.2 On-line services

While maintaining local collections of databases and software is possible and may be necessary for some research groups, due to the size and growth rate of current biological databases significant efforts are required to keep up with the regular updates. Databanks such as EMBL and GenBank distribute new database entries on a daily basis; however, only a few centres can afford the effort required to integrate these new sequences each day with the main tools for database searching and analysis. The costs of human resources needed to maintain local database collections can quickly exceed the costs of expensive hardware and software. While large laboratories may be able to cope with these problems and even afford their own informatics staff, many researchers in small institutes will find it more cost-effective to use external computing services instead.

A number of organizations are now offering comprehensive informatics services to molecular biologists.

3.2.1 EMBnet services

European telecommunication costs and performance, especially across national boundaries, can make direct interactive access to one central international computing service prohibitive for many users. As a consequence, the European Molecular Biology Network (EMBnet) project was initiated in 1988 aiming at establishing an infrastructure of collaborating nodes in most European countries, each node providing biocomputing services, including user support and training, to their national academic and commercial user community. A list of EMBnet national nodes currently in operation is provided in *Table 2*. A national node typically offers access to daily updated sequence databases, a collection of other molecular biological databases, comprehensive sequence analysis software such as the GCG suite, and access to international computer networks and bulletin boards. Many nodes offer services beyond that, often resulting from, and taking into account, local research interests. A brochure which describes the EMBnet project in more detail can be obtained from the national nodes listed in *Table 2*.

3.2.2 Other on-line services

While EMBnet is the predominant on-line service provider within Europe, similar systems are also available in other continents and countries, and some of them are shown in *Table 2*. Many of them offer their services for free or charge a nominal fee, others are commercial enterprises with corresponding price structures. Personal choice will depend on various factors such as local availability, usage fees, the support offered and, most importantly, a close correspondence of the services offered with personal needs.

Table 2. On-line service providers

Name	**Contact person, fax number e-mail address**
EMBnet nodes	
EMBnet node Belgium	R. Herzog, +32–2–6509762 rherzog@ulb.ac.be
EMBnet node Denmark	H. Møller, +45–86–131160 hum@biobase.aau.dk
EMBnet node Finland	R. Harper, +358–0–4572302 harper@csc.fi
EMBnet node France	P. Dessen, +33–1–69333013 dessen@coli.polytechnique.fr
EMBnet node Germany	W. Chen, +49–6221–422333 dok419@genius.embnet.dkfz-heidelberg.de
EMBnet node Greece	B. Savakis, +30–81–231308 savakis@myia.imbb.forth.gr
EMBnet node Israel	L. Esterman, +972–8–344113 lsestern@weizmann.weizmann.ac.il
EMBnet node Italy	M. Attimonelli, +39–80–484467 attimonelli@mvx36.csata.it
EMBnet node Netherlands	J. Noordik, +31–80–652977 noordik@caos.kun.nl
EMBnet node Norway	R. Lopez, +47–22694130 rodrigol@biomed.uio.no
EMBnet node Spain	L. Pezzi, +341–585–4506 lpezzi@cnbvx3.cnb.uam.es
EMBnet node Sweden	P. Gad, +46–18–551759 gad@perrier.embnet.se
EMBnet node Switzerland	R. Dölz, +41–61–2672078 doelz@comp.bioz.unibas.ch
EMBnet node United Kingdom	A. Bleasby, +44–925–603100 bleasby@daresbury.ac.uk
Other services	
UK Human Genome Mapping Project (HGMP) Resource Centre	F. Rysavy, +44–81–4231275 frysavy@crc.ac.uk
International Centre for Genetic Engineering and Biotechnology	S. Pongor, +39–40–226555 pongor@genes.icgeb.trieste.it
MIPS Max–Planck-Institut für Biochemie	H. W. Mewes, +49–89–8578–2655 mewes@ehpmic.mips.biochem.mpg.de
Brazilian Molecular Biology and Biotechnology Network	G. Neshich, +55–61–273–8350 bbrc@cenargen.embrapa.br
European Data Resource for Human Genome Analysis	O. Ritter, +49–6221–401271 dok261@cvx12.dkfz-heidelberg.de
BioTechNet	BioTechNet, +1–508–655–9910
Australian National Genomic Information Service (ANGIS)	A. Reisner, +xx–2–692–3847 reisner@angis.su.oz.au
BIONET Timesharing service	D. Kristofferson, +1–415–962–7302 kristoff@net.bio.net
MBCRR Harvard, Boston	T. Smith, phone +1–617–732–3746 tsmith@mbcrr.harvard.edu

3.3 Electronic mail servers

The ability to send electronic mail (e-mail) is probably the best single reason for connecting to international computer networks. Electronic mail not only allows a researcher to communicate with colleagues in other parts of the world faster, more efficiently, and more conveniently than with normal mail, it also enables the establishment of *electronic mail servers* which respond to requests sent across the networks from remote sites. An e-mail server is simply a computer program that—in a fully automated process—waits for incoming messages, interprets these messages, and replies according to the requested action.

Recent years have seen an explosion of the number of e-mail servers. Servers that enable access to molecular sequence databases have been complemented by new servers that perform specialized data analysis tasks, often using software and hardware not available elsewhere. These servers are generally easy to use, and no charges occur besides the general costs of using a computer network. A drawback is that popular servers are often flooded by requests and response times may vary significantly as a function of the load level.

3.3.1 Using electronic mail servers

If you know how to send electronic mail on your system you know how to use e-mail servers. E-mail servers are simple computer programs and are limited in what they can understand. Thus messages to e-mail servers, therefore, have to be correctly formatted. Unfortunately, the format and options accepted differ from server to server. Almost all servers, however, recognize the command HELP and reply to it with a detailed description of the service, including instructions on how to submit requests. The first e-mail message to a new server should therefore always be the word HELP on a single line of itself.

The details of how mail is sent from a particular computer system vary from platform to platform. *Figure 2* shows how a message is sent to `netserv@embl-heidelberg.de` from a UNIX computer system. The first line requests general instructions while the next two lines retrieve single entries from the EMBL and SWISS-PROT databases. The fourth line is a request for the complete eukaryotic promoter database (EPD). Sending mail from other systems should be equally simple.

The occasional user may feel overwhelmed by the sheer number of servers available and the diversity of options and formatting rules. However, tools exist which facilitate the use of e-mail servers by providing menus of servers and options, prompting the user interactively for required information and sending off correctly formatted requests to the server of choice (e.g. see ref. 10).

```
% mail  netserv@embl-heidelberg.de
Subject: Test
help
get  nuc:x13999
get  prot:wap_mouse
get  epd:epd.dat
.
EOT
%
New mail for fuchs@felix has arrived:
----
Date: Fri, 15 Oct 1993 16:14:49 +0100
From: NETSERV@EMBL-Heidelberg.DE
Subject: Reply to: HELP
To: fuchs@felix.embl-heidelberg.de

HELP [GENERAL]

...more...
%
```

Figure 2. Sending an electronic mail server request. User input is shown in bold.

3.3.2 Database servers

A variety of electronic mail servers enable researchers to download complete databases as well as individual database entries. Most interestingly, some servers give access to daily updates of their data collections. For instance, when a nucleotide sequence is published in a journal, it will normally be possible to retrieve the corresponding database entry from netserv@embl-heidelberg. de or retrieve@ncbi.nlm.nih.gov at the same time as the paper appears, using the accession number cited in the publication.

Table 3 lists the major electronic mail database servers and a variety of smaller sites. A comprehensive list can be found in ref. 11.

3.3.3 Sequence analysis servers

Recently, traditional database e-mail servers have been supplemented by a number of new servers which permit specific sequence analysis tasks to be performed at remote sites. These include investigations such as: similarity searches, which need access to most recent data collections; tasks that require unusual hardware, such as massively parallel computers; or specialized software not available elsewhere, such as gene or protein structure prediction programs based on neural net approaches.

Table 4 gives an overview of existing sequence analysis servers while more detailed descriptions can be found in ref. 11. Remember that most of these servers return extensive documentation when a user sends an electronic mail message with a line containing the single word HELP.

Table 3. Electronic mail database servers

Electronic mail address	Databases
netserv@embl-heidelberg de netserve@ebi.ac.uk	General data repository, complete databases and single entries, incl. EMBL and SWISS-PROT daily updates
retrieve@ncbi.nlm.nih.gov	Entries from various databases incl. GenBank and EMBL daily updates
fileserv@nbrf.georgetown.edu	Entries from PIR protein sequence database
fileserv@pb1.pdb.bnl.gov	Entries from Brookhaven protein structure database (PDB)
gene-server@bchs.uh.edu	Entries from PIR protein sequence database
server@rdp.life.uiuc.edu	Data from Ribosomal Database Project
est–report@ncbi.nlm.nih.gov	Entries from dbEST expressed sequence tag database
flat-netserv@smlab.eg.gunma-u.ac.jp	Entries from GenBank, EMBL, SWISS-PROT, PIR
dbget@genome.ad.jp	Entries from various databases, including GenBank daily updates

Table 4. Electronic mail sequence analysis servers

Name	Task
blitz@embl-heidelberg.de	Exhaustive protein/nucleotide database searches on massively parallel computer
dflash@watson.ibm.com	Exhaustive protein database searches
bioscan@cs.unc.edu	Protein/nucleotide database searches on massively parallel computer
blast@ncbi.nlm.nih.gov	Protein/nucleotide database searches
blast@genome.ad.jp	Protein/nucleotide database searches
fasta@ebi.ac.uk	Protein/nucleotide database searches
fasta@genome.ad.jp	Protein/nucleotide database searches
quick@embl-heidelberg.de	Nucleotide database searches
fileserv@nbrf.georgetown.edu	Protein database searches
blocks@howard.fhcrc.org	Sequences comparisons against database of protein blocks
motif@genome.ad.jp	Protein sequence motif identification
pythia@anl.gov	Identification of repetitive elements
geneid@darwin.bu.edu	Exons structure/gene prediction
grail@ornl.gov	Coding region/gene prediction
genmark@ford.gatech.edu	Gene prediction
netgene@virus.fki.dth.dk	Splice-site prediction in vertebrates
predictprotein@embl-heidelberg.de	Secondary structure prediction
nnpredict@celeste.ucsf.edu	Secondary structure prediction
cbrg@inf.ethz.ch	General sequence analysis tools

3.4 Anonymous FTP servers

The file transfer protocol (FTP) is the standard mechanism for exchanging files between hosts on the Internet. A site providing access to information in this way is referred to as an *FTP server*. In general, it does not matter which particular computer systems are involved in an FTP communication; data exchange between a VMS and an MS–DOS computer is as seamless as between two UNIX systems, as long as both machines are on the Internet. The word 'anonymous' indicates that access to a host is possible for everyone, without the need of having a user account on the host computer. Anonymous FTP is thus a simple way of making information available to a world-wide user community.

A major disadvantage of FTP is its limited functionality. In essence, it is a means for transferring a file between one computer system and another. Unlike electronic mail servers, database searches or submissions of sequence analysis jobs are not possible this way. Despite that, the popularity of anonymous FTP servers has outgrown that of electronic mail servers in recent years, mainly due to the immediate interaction that takes place which seems to give users a greater control over their activities. Also, communication is often faster than with electronic mail servers, because popular mail servers can develop backlogs of requests, resulting in response times of hours and longer. As a result, several thousand anonymous FTP servers are now accessible on the Internet, many of them storing information and data relevant to biologists.

3.4.1 Using FTP

All that is required to use FTP is the FTP client software on the local computer system. *Figure 4* shows a typical FTP session on a UNIX workstation. It illustrates that normally only a small number of commands are used. Even though the commands given here are common and found for most FTP clients, their actual names may differ in some cases.

First, open an FTP session by issuing the **ftp** command with the Internet name of the desired host. In *Figure 4A* we connect to the host `ftp.embl-heidelberg.de`. If the host is available and the connection can be established, the user is prompted for a user name, and because the server is an anonymous FTP server, the reply is **anonymous**. The next question then is for a password, and one should reply by giving his or her full electronic mail address. Now we are connected and free to explore the information provided by this host.

Most FTP servers follow the UNIX syntax for naming their directory structure, although there are exceptions to this. In many cases, there will be a README file at the top level, explaining the peculiarities of a system. Users are strongly advised to download and read this document. The entry point into the data archive is often a directory called `/pub`, and the data

A

```
% ftp ftp.embl-heidelberg.de
Connected to felix.EMBL-Heidelberg.DE.
220 felix FTP server (Version 2.1WU(1) Fri May 14 13:20:24 MET 1993)
ready.
Name (ftp.embl-heidelberg.de:fuchs): anonymous
331 Guest login ok, send your complete e-mail address as password.
Password:
230-
230- >>> Welcome to the EMBL molecular biology ftp archive! <<<
230-
230- The local time is Fri Oct 15 12:53:21 1993.
230-
230 Guest login ok, access restrictions apply.
ftp> dir
200 PORT command successful.
150 Opening ASCII mode data connection for /bin/ls.
total 1067
-rw-r--r--  1 root     ftp       870445 Oct 14 22:10 INDEX
-rw-r--r--  1 root     ftp       186709 Oct 14 22:28 NEWFILES
drwxr-xr-x  2 root     ftp          512 Feb  5  1993 bin
drwxr-xr-x  3 root     ftp          512 Feb  5  1993 etc
drwxr-xr-x  2 root     0           8192 Feb  5  1993 lost+found
drwxr-xr-x  9 ftpadmin ftp          512 Jul 16 13:29 pub
-rw-r--r--  1 root     0              0 Sep 29 17:47 trace.dump
226 Transfer complete.
446 bytes received in 0.13 seconds (3.4 Kbytes/s)
ftp> get NEWFILES
200 PORT command successful.
150 Opening ASCII mode data connection for NEWFILES (186709 bytes).
226 Transfer complete.
local: NEWFILES remote: NEWFILES
188778 bytes received in 1.4 seconds (1.3e+02 Kbytes/s)
```

B

```
ftp> cd /pub/databases/embl/new
250-In this directory you can find the latest additions and updates to
250-the EMBL nucleotide sequence database.
250-
250-Files in this directory ending with .Z have been compressed
250-using the UNIX compress tool. Transfer these files in
250-BINARY mode to your system.
250-
250 CWD command successful.
ftp> dir
200 PORT command successful.
150 Opening ASCII mode data connection for /bin/ls.
total 11489
-rw-r--r--  1 ftpadmin ftp          250 Nov  9  1992 .message
-rwxr-xr-x  1 ftpadmin ftp      1236454 Oct  3 03:12 931003.dat.Z
-rwxr-xr-x  1 ftpadmin ftp      2212285 Oct 10 04:14 931010.dat.Z
drwxr-xr-x  2 ftpadmin ftp         8192 Apr 25  1991 lost+found
-rwxr-xr-x  1 ftpadmin ftp       404966 Oct 10 04:14 newacnumber.ndx
-rwxr-xr-x  1 ftpadmin ftp      1346730 Oct 10 04:14 newauthor.ndx
-rwxr-xr-x  1 ftpadmin ftp       361295 Oct 10 04:14 newcitation.ndx
-rwxr-xr-x  1 ftpadmin ftp       177146 Oct 10 04:14 newentries.ndx
-rwxr-xr-x  1 ftpadmin ftp       309877 Oct 10 04:14 newkeyword.ndx
-rwxr-xr-x  1 ftpadmin ftp       382400 Oct 10 04:14 newshortdir.ndx
-rwxr-xr-x  1 ftpadmin ftp       253288 Oct 10 04:14 newspecies.ndx
226 Transfer complete.
1043 bytes received in 0.12 seconds (8.7 Kbytes/s)
ftp> binary
200 Type set to I.
ftp> mget 9310*
200 PORT command successful.
150 Opening BINARY mode data connection for 931003.dat.Z (1236454
bytes).
226 Transfer complete.
local: 931003.dat.Z remote: 931003.dat.Z
1236454 bytes received in 11 seconds (1.1e+02 Kbytes/s)
200 PORT command successful.
150 Opening BINARY mode data connection for 931010.dat.Z (2212285
bytes).
226 Transfer complete.
local: 931010.dat.Z remote: 931010.dat.Z
2212285 bytes received in 24 seconds (90 Kbytes/s)
ftp> quit
221 Goodbye.
%
```

Figure 3. An example FTP session. User input is shown in bold.

available is organized into subdirectories of /pub. Many anonymous FTP servers provide in their top-level directory a listing of all files available on the server or a listing of all files new or updated during the last few days. It is generally more efficient to download these files and examine them off-line than to interactively browse through an expanded directory hierarchy looking for new data. This is shown in *Figure 3A* where we download a file called NEWFILES containing latest changes.

To change between different directories, the command **cd** is used. In the example shown in *Figure 3B* we move to the directory that holds the latest EMBL database entries. The **dir** command will make the remote host send a listing of all files in the current directory including their sizes. Many hosts have README files in each directory which give a detailed explanation of the content of the directory and the files it holds. Individual files can be downloaded with the **get** command. The downloading of multiple files can be simplified by issuing the **mget** command using filenames with wildcard characters.

Many FTP servers store data in compressed form, to save disk space and to reduce download times. Most often, the UNIX compress utility is used for shrinking files, and files treated this way can be identified by the .Z file name extension (for example, 931010.dat.Z in *Figure 3B*). It is important to use the right mode for downloading these files. By default, files are downloaded in *ascii mode* which is appropriate for text files. For binary files such as .Z files or executables such as MS–DOS .EXE files, *binary mode* must be used instead. To toggle between these modes, simply issue the **ascii** or **binary** command.

FTP connections should be properly closed by sending the **quit** command to the remote host at the end of the session.

3.4.2 Molecular biology FTP servers

The recent past has seen the establishment of an increasing number of anonymous FTP archives that make molecular biological data and information available. Many servers provide only a limited set of information, such as a specialized database or a particular software package, but there are a few 'one-stop shopping' sites with comprehensive collections of databases and other information. Ftp.ebi.ac.uk and ncbi.nlm.nih.gov, for instance, provide regular updates to databases such as EMBL/GenBank or SWISS-PROT, so that it is possible to update local copies of these databases across the network. Some genome centres, such as the Institute for Genome Research (TIGR) or Genethon Paris, also make their sequence and mapping data available via the FTP mechanism. *Table 5* lists the major servers plus some of the more important smaller FTP sites. A more comprehensive list can be found in ref. 12.

Table 5. Molecular biology anonymous FTP servers

Name	Internet address
EMBL Data Library	ftp.ebi.ac.uk
US National Center for Biotechnology Information (NCBI)	ncbi.nlm.nih.gov
Genome Data Base (GDB)	ftp.gdb.org
Protein Data Bank (PDB)	ftp.pdb.bnl.gov
National Institute of Genetics Japan	ftp.nig.ac.jp
Indiana University Department of Biology	ftp.bio.indiana.edu
Japanese Human Genome Project server	ftp.genome.ad.jp
University of Geneva, ExPASy server	expasy.hcuge.ch
The Institute for Genome Research (TIGR)	ftp.tigr.org
International Centre for Genetic Engineering and Biotechnology (ICGEB) Trieste	ftp.icgeb.trieste.it
Ribosomal Database Project (RDP)	rdp.life.uiuc.edu
SERC Daresbury	s-crim1.dl.ac.uk
Biozentrum Basel, EMBnet national node	bioftp.unibas.ch

3.5 Gopher servers

Although computers have reached most areas of modern biology, many biologists are still less happy dealing with computers than with one of their traditional laboratory devices. Friendly graphical user interfaces on desktop computers can help much to lower the barrier. While programs such as FTP are not particularly difficult to use most life scientists prefer to stay in the point and click environment they are accustomed to from their Macintosh and MS Windows computers.

Computer scientists have recently developed some new solutions that hide computer networks and their complexity behind user-friendly graphical interfaces which allow even the computer-illiterate scientist to explore remote information resources and download relevant data files easily. Probably the most successful of these tools is Gopher (13) which was developed at the University of Minnesota as a distributed document browsing/delivery system based on a server–client communication model. Gopher client software is available for most computer platforms, including the Macintosh and MS-DOS systems. It allows a user to access information residing on multiple hosts on a network without a need for her to know where the data are physically located. Within less than two years of its appearance, several hundred Gopher servers have been established all over the world covering almost all areas of science.

3.5.1 Using Gopher

Figure 4 shows a typical Gopher session using the UNIX line-oriented Gopher client. Assume we are looking for information on hybridomas available from ATCC. The first thing to do in a Gopher session is to connect to

A

```
% gopher ftp.embl-heidelberg.de 70
                         Internet Gopher Information Client 2.0 pl6
                         Root gopher server: ftp.embl-heidelberg.de

        1.  About Gopher  [15Jun92, 2kb].
    --> 2.  About This Resource  [15Jun92, 2kb].
        3.  EMBnet BioInformation Resource EMBL/
```

B

```
EMBnet BioInformation Resource EMBL
------------------------------------

This is a BioGopher server running as part of the EMBnet BioInformation
Resources network. The server is maintained by the EMBL Data Library at the
European Molecular Biology Laboratory in Heidelberg, Germany.

It provides access to a variety of data collections and other information
in molecular biology, to free software for MS-DOS, VAX/VMS, Unix and
Macintosh systems, and to other gophers and biogophers.

[...rest deleted...]
```

C

```
                         Internet Gopher Information Client 2.0 pl6
                         Root gopher server: ftp.embl-heidelberg.de

        1.  About Gopher  [15Jun92, 2kb].
        2.  About This Resource  [15Jun92, 2kb].
    --> 3.  EMBnet BioInformation Resource EMBL/
```

D

```
                         Internet Gopher Information Client 2.0 pl6
                            EMBnet BioInformation Resource EMBL

        1.  About EMBnet/
        2.  Databases/
        3.  Hints/
    --> 4.  Other EMBnet Hosts and Biological Sources/
        5.  Other Information Resources/
        6.  Software/
```

E

```
                   Internet Gopher Information Client 2.0 pl6
                   Other EMBnet Hosts and Biological Sources

     1.  Search Biological Gopher-Space - BOING (Bio Oriented INternet G..
<?>
     2.  Global Biological Information Servers/
     3.  Global Biological Information Servers by Topic/
 --> 4.  ATCC Databases/
     5.  Arabidopsis AAtDB Gopher Server /
     6.  BIOSCI Gopher/
     7.  BioGopher University Kaiserslautern (Germany)/
     8.  BioInformatics gopher at ANU/
     9.  Biodiversity and Biological Collections at Harvard/
     10. Biology FTP Servers/
     11. Biology subject tree in Gopher/
     12. Computational Biology (Bookreading via GOPHER)/
     13. Databases and Information Resources at the Genome Database (GDB)/
     14. EMBnet BioBox Finland/
     15. EMBnet BioInformation Resource (France)/
     16. EMBnet BioInformation Resource (Greece)/
     17. EMBnet BioInformation Resource (Israel)/
     18. EMBnet BioInformation Resource (Norway)/
```

F

```
                       Internet Gopher Information Client 2.0 pl6
                                    ATCC Databases
    --> 1.  ATCC Catalogs (courtesy of Johns Hopkins University)/
        2.  About this Gopher.
        3.  E-mail and Directory Services/
        4.  Gopher Tunnels (To Other Gopher and WAIS Sites)/
        5.  Grants and Research Information/
        6.  Hybridoma Data Bank (courtesy of RIKEN)/
        7.  Molecular Biology Databases (non-ATCC)/
        8.  Phone Books (provided by University of Minnesota)/
        9.  Weather and Area Information/
        10. Welchlab at Johns Hopkins University/
```

Figure 4. An example Gopher session.

a *root* server. The address of such a root server is the only thing one needs to know about computer networks. The example in *Figure 4A* shows a root connection to the server running at EMBL Heidelberg. The server replies with a menu of options. By choosing option 2 we look in *Figure 4B* at a file which describes the EMBL Gopher information resource. In *Figure 4C*, we choose option 3 to get down to the actual information archive. The new menu shown in *Figure 4D* gives us the opportunity to look at various databases, to download software, or to retrieve other information. We choose option 4 to see which other biological information resources can be reached from here and receive the menu shown in *Figure 4E*. From here option 4 leads us to the ATCC databases, and the new menu *Figure 4F* lets us open the ATCC catalogues or search the Hybridoma Data Bank for the data we are interested in.

Note that throughout this whole procedure the knowledge where data is physically located on the Internet is hidden from the user. In fact, the menu shown in *Figure 4E* is supplied by a Gopher server running at the NIH in Bethesda, USA, for example. Network usage with Gopher is completely transparent and there is no need for the researcher to remember Internet addresses, formats and options, or command line syntaxes.

3.5.2 Biological Gopher servers

More than a hundred biological Gopher servers (BioGophers) are now available, with information ranging from standard sequence databases, genome databases, or physical and genetic maps, to catalogues of strain and stock centres, toxicological data, oceanographic information, addresses of scientists, and so on. Increasingly, this mechanism is utilized by genome centres for making their data widely available to the general user community. It is now possible with Gopher, for example, to look at the *Drosophila* genome database Flybase, to query the *Arabidopsis thaliana* database AAtDB, or to browse information on the *Saccharomyces* genome.

In contrast to anonymous FTP or electronic mail servers, the exact choice of a server to connect to is not very important with Gopher. The many existing links between BioGophers enable users to seamlessly navigate from one server to the other without even knowing that they are switching hosts. To get started, some good root servers which provide plenty of links to other BioGophers are indicated in *Table 6*. A comprehensive list of all Gopher servers world-wide is maintained at the 'Gopher home' `gopher.micro.umn.edu`.

3.6 Other tools for network information retrieval

The importance of network information retrieval is perhaps best illustrated by the rapidly growing number of software tools developed for this purpose (14). This section gives a cursory overview of some general tools that have been applied to molecular biology, and it describes some systems especially developed for retrieval of biological data across computer networks.

3.6.1 WAIS and the World-Wide Web

WAIS, an acronym for Wide Area Information System, is an attempt to simplify access to remote databases by hiding the network aspects from the users and by providing a natural language query interface (15). WAIS is based on the client–server approach. Clients are available for most hardware platforms and include simple line-oriented interfaces as well as sophisticated graphical user interfaces. Also, through the development of a gateway between WAIS and Gopher, WAIS searches can be performed from within standard Gopher client software. The databases available for searching are presented to the user as so-called *sources*. These sources may be local or remote, but the user does not need to know anything about a source's location because the details of how to connect to and how to access a source are handled transparently by the client. A distinctive feature of WAIS is its ability to understand queries formulated in standard English such as 'what is the role of tyrosine kinases in intracellular signal transduction'. Thus, there is no need for a user to learn an awkward query language. The result of a WAIS search is a ranked list of documents that contain information relevant to the query sentence.

Table 6. Molecular biology Gopher servers

Name	Internet address
EMBnet BioInformation Resource Switzerland	bioftp.unibas.ch
IUBio Archive	ftp.bio.indiana.edu
EMBnet BioInformation Resource EMBL	gopher.embl-heidelberg.de
Yeast Genome Information Server	genome-gopher.stanford.edu
Computational Biology (Johns Hopkins U.)	gopher.gdb.org
BioInformatics at Australian Natl. University	life.anu.edu.au

Today, many biological sources are available for WAIS searches, including sequence and mapping databases. A (WAIS-searchable) directory of all sources is maintained at `quake.think.com` and at `cnidr.org` from which WAIS client software is also available by anonymous FTP. The usefulness of WAIS searches in conjunction with biological databases is questionable, however. WAIS was primarily developed and optimized for free-text searches of natural language text documents such as financial reports or newspaper articles, and the ranking of search hits is largely dependent on the redundancy of words in the database. Databases with low redundancy such as biological data collections often fool the WAIS ranking system and let interesting hits disappear among spurious hits.

Perhaps the most ambitious approach to network information retrieval these days is the *World-Wide Web* (WWW). Originally designed and developed by scientists at CERN in Geneva, WWW adopts the hypertext model on a world-wide scale in a client–server communication system (16). A key element of WWW is a special text mark-up language which allows the embedding of links to other documents located at different sites into a document. So when viewing a text document with a graphically oriented WWW client program, the user can simply click on a part of the text and is taken immediately to a related document with additional information about the selected text element. This linking mechanism opens up interesting new ways of representing relationships between different data items and data collections.

WWW client software provides gateways to Gopher and WAIS and can thus be used as a general network information retrieval tool. However, genuine WWW servers for biological information are still rare, mainly because the preparation of documents with embedded links is a tedious process which requires significant human effort. A list of WWW servers that cover biological topics is maintained at Harvard University (http://golgi.harvard.edu/biopages.list). WWW client software is available from info.cern.ch and ftp.ncsa.uiuc. edu. These servers also offer 'home pages' with links to other information resources, including those of interest to biologists.

3.6.2 Biological client–server network retrieval systems

While access to biological network information resources is currently mainly based on general protocols and software such as FTP, Gopher, or WWW, a few specialized client–server applications have been developed recently for access to sequence and mapping databases. Only two particularly interesting examples are described here, but we can expect that many more similar tools will be created in the near future.

Network Entrez is a derivative of NCBI's Entrez program. While Entrez allows access to sequence and literature data on CD–ROM, Network Entrez uses a database server running on NCBI's computer facilities as a back-end. The user interface is exactly the same, and the fact that Network Entrez accesses remote databases across the Internet is completely hidden from the user. This way it is possible not only to search regular database releases as provided on CD–ROM but to query daily updated data collections containing most recent information. With high-speed data links, such as those within the USA, access to the remote NCBI databases can be even faster than to a local CD–ROM drive, while on transatlantic connections, for instance between Europe and the USA, response time is often noticeably affected by transmission delays.

GDB/Accessor is an Apple Macintosh program developed by Cold Spring Harbor Laboratories and the genome database (GDB) and provides a graphical front-end to the GDB relational database of human mapping information. Various query screens allow the easy construction of highly complex queries which are then automatically translated into the corresponding SQL queries understood by the relational database management system running at GDB

Table 7. Major FTP servers for molecular biological software

Address	Software available
ftp.ebi.ac.uk	Various DOS, Mac, UNIX, and VMS software
ftp.bio.indiana.edu	Various DOS, Mac, UNIX and VMS software
ftp.bchs.uh.edu	Various DOS, Mac, UNIX, and VMS software
ftp.nig.ac.jp	Various DOS, Mac, and UNIX software
sunbcd.weizmann.ac.il	'Mirror' of the EBI software archive

or their world-wide satellites. Queries are sent to the nearest GDB node and within a short while results are returned and graphically displayed.

Programs such as Network Entrez or GDB/Accessor are excellent examples for a new generation of database query tools which provide simple graphical user interfaces that allow the easy construction of queries without the need to learn a special query language, and which make network access to remote databases transparent to the user.

4. Software archives

Over the last few years several organizations have invested significant efforts into the establishment of central archives or repositories of free molecular biological software. These 'software supermarkets' are particularly interesting for research groups that do not want or can not afford to invest into expensive commercial software packages. The software available from these archives covers most application areas of molecular biology, ranging from restriction mapping to gene modelling, or from enzyme kinetics calculation to linkage analysis. Besides these major places there is a large number of sites that provide access to locally developed software packages.

The most convenient way of accessing these archives is by anonymous FTP (see section 3.4) or Gopher (see section 3.5). *Table 7* lists the major FTP software archives on the Internet, while a more complete directory of large and smaller sites is available in refs 11 and 12. Those without the possibility to use FTP can download software via electronic mail from `netserv@ ebi.ac.uk` or `gene-server@bchs.uh.edu`. While transferring binary executable files is easy using the FTP protocols, this is not possible with electronic mail. Binary files must be converted into some character-based format that allows transport by electronic mail ('uuencoding', 'binhexing'). One should therefore carefully study the help files provided by the electronic mail software servers that give detailed instructions describing the decoding process and the programs required.

5. Conclusions

The importance of biological databases is ever growing and efficient methods for accessing and querying these information resources are crucial to the work of any biologist.

Unfortunately, the costs of local database maintenance are high in view of the growing number and sizes of databases. Maintaining most recent local copies of several major databases may be prohibitively expensive for most small laboratories. CD–ROM technology and CD–ROM-based query software can help remove many of these obstacles, but are mainly limited by reduced speed and prolonged release cycles. For the foreseeable future, CD–ROMs will probably form a core data set which will be analysed first and which will be supplemented by approaches based on direct computer communication.

Access to latest data is increasingly important and only possible via computer networks. On-line services such as offered by EMBnet nodes, provided regularly with new data from the major databanks, are an attractive way of utilizing a variety of complete databases and the necessary analysis software.

The future trend will be towards a distributed database environment in which a scientist uses his local desktop computer to directly connect to databases spread around the world. Advances in network technology, together with the development of new tools for information discovery and retrieval, will allow a world-wide information network providing access to databases made available directly by the producers.

With sophisticated server–client communication models and graphical user interfaces that hide the network aspect and make use of remote databases transparent, for the biologist using databases will be as easy as never before.

References

1. von Heijne, G. (1987). *Sequence analysis in molecular biology*. Academic Press, London.
2. Keen, G., Redgrave, G., Lawton, J., Cinkosky, M., Mishra, S., Fickett, J., *et al.* (1992). *Math. Comput. Modelling*, **16**, 91.
3. Fuchs, R. and Cameron, G. N. (1991). *Prog. Biophys. Mol. Biol.*, **56**, 215.
4. Damerval, T. and Dessen, P. (1992). *Biofuture*, **53**, 3.
5. Bishop, M., Ginsburg, M., Rawlings, C. J., and Wakeford, R. (1987). In *Nucleic acid and protein sequence analysis: a practical approach* (ed. M. J. Bishop and C. J. Rawlings), pp. 83–113. IRL Press, Oxford.
6. Fuchs, R., Rice, P., and Cameron, G. N. (1992). *TIBTECH*, **10**, 61.
7. Doolittle, R. F. (ed.) (1990). *Methods in enzymology*, Vol. 183. Academic Press, London.
8. Coulson, A. (1993). *Trends Biotechnol.*, **11**, 223.
9. Etzold, T. and Argos, P. (1993). *Comput. Appl. Biosci.*, **9**, 49.

10. Fuchs, R. (1993). MSU—Mail Server Utility. Available electronically from the EMBL anonymous FTP server as file://ftp.embl-heidelberg.de/pub/software/unix/msu.tar.Z and file://ftp.embl-heidelberg.de/pub/software/vax/msu.uue.
11. Bairoch, A. (1993). List of molecular biology e-mail servers. Published electronically on the Internet. Available from the EMBL anonymous FTP server as file://ftp.embl-heidelberg.de/pub/doc/serv_ema.txt.
12. Bairoch, A. (1993). List of molecular biology FTP servers for databases and software. Published electronically on the Internet. Available from the EMBL anonymous FTP server as file://ftp.embl-heidelberg.de/pub/doc/serv_ftp.txt.
13. McCahill, M. (1992). *ConneXions—The Interoperability Report*, **6**, 10.
14. Schwartz, M. F., Emtage, A., Kahle, B., and Neuman, B. C. (1992). *Computing Systems*, **5** (published as a preprint in file://quake.think.com/wais/wais-discussion/issue-56.text).
15. Kahle, B. and Medlar, A. (1991). *ConneXions—The Interoperability Report*, **5**, 2.
16. Berners-Lee, T., Cailliau, R., Groff, J., and Pollermann, B. (1992). *Electronic networking: research, applications and policy*, **2**, 52.

7

Long-range restriction mapping

WENDY BICKMORE

1. Introduction

Mapping large and complex genomes relies on techniques that permit the analysis and manipulation of long stretches of DNA, hundreds to thousands of kilobases in length. This has been achieved largely through the developments of pulsed field gel electrophoresis (PFGE)—allowing the resolution of large DNA fragments, and yeast artificial chromosome vectors (YACs) for cloning large DNA fragments. Being able to resolve very large fragments of DNA also necessitates being able to restrict DNA into suitably long lengths. This can be accomplished with restriction enzymes with large, and therefore rare recognition sequences. In vertebrates, large restriction fragments can also be generated using restriction enzymes with CpG dinucleotides in their recognition sequences (this also applies to plant DNAs and includes enzymes with CpXpGs in their recognition sites). The principle cleavage sites for CpG recognizing enzymes in vertebrate genomic DNA, as well as providing convenient landmarks for the construction of long-range restriction maps, also mark out the position of many of the genes—critically important in organisms where only a few per cent of the DNA has a coding capacity. This chapter aims to demonstrate how to create and interpret long-range restriction maps, primarily, of mammalian genomes.

2. Preparing intact large DNA molecules

During the normal preparation of DNA in solution, shear forces break the long thin molecules down to sizes less than 500 kb. To restrict and analyse defined DNA molecules considerably larger than this, DNA must be protected from shear forces during its preparation. This is done by encapsulating cells, prior to lysis, in agarose beads (1) or blocks ('plugs'). For ease of handling the latter method is the most commonly used (*Protocol 1*). Methods have been described for preparing high molecular weight DNA from a wide range of higher eukaryotes. These include mammalian and non-mammalian vertebrates (2), and plant DNAs (3, 4). Protocols for preparing high molecular weight mammalian genomic DNA and YAC clone DNA are described in *Protocols 1* and *2*, respectively.

2.1 Embedding mammalian cells in agarose plugs

DNAs from many different species and from diverse tissue types have been prepared for PFGE by embedding cells in agarose plugs. The latter have included blood, solid tissues, cell lines, and even frozen tumour samples (5). The source of biological starting material can influence the way in which long-range maps are interpreted and this will be discussed in section 11. The agarose used for making the plugs must be of a low gelling temperature variety, be nuclease-free, and of a high enough purity so that it will not inhibit the action of restriction enzymes. We routinely use BRL Ultrapure LMP agarose, though there are many other brands that are satisfactory; e.g. InCert agarose (FMC Bioproducts). To ensure even distribution of DNA throughout the plugs, cells must be in single cell suspension. Cells grown as monolayer cultures should be harvested by trypsinization and any cell clumps broken up with a fine-tipped pastette. Fresh tissue samples should be homogenized in ice-cold PBS and filtered through muslin to remove connective tissue. Frozen tissue is ground at −70°C, resuspended in cold PBS, and similarly filtered through muslin. When fresh blood is used as the source of starting material, red cells are first lysed in three volumes of lysis buffer (155 mM NH_4Cl, 10 mM $KHCO_3$, 0.1 mM EDTA pH 7.4). White cells are then spun out at 250 *g* for 5 min at 4°C, and washed once more in lysis buffer before embedding in agarose.

Protocol 1. Embedding mammalian DNA in agarose plugs[a]

Equipment and reagents

- NDS: 0.5 M EDTA, 10 mM Tris–HCl, 1% (w/v) lauryl sarcosine pH 9.5 (use NaOH pellets to maintain the pH whilst dissolving the EDTA; autoclave before adding lauryl sarcosine)
- LMP ultrapure agarose (BRL)
- Plug mould (Bio-Rad)
- Proteinase K (BCL)

Method

1. Harvest the cells as appropriate (see text) and pellet by centrifugation at 4°C (250 *g*, 5 min) and wash them three times in ice-cold PBS.
2. Count the cells, re-pellet them, and resuspend in PBS at 5×10^7 cells/ml.
3. Add an equal volume of molten 1% (w/v) LMP ultrapure agarose (BRL) made up in PBS and maintained at 42°C. Mix the molten agarose and cells well and dispense 100 µl aliquots into plastic moulds (0.2 × 0.5 × 1.0 cm).

4. Leave the moulds in the fridge until the agarose is set, then push the plugs gently out of the moulds and into a sufficient volume of NDS to cover them well.
5. Add proteinase K to 0.5 mg/ml. Incubate at 50°C for 24 h.
6. Replace with fresh NDS and proteinase K and continue the incubation for a further 48 h.
7. Store the plugs at 4°C in fresh NDS. They will keep for several years like this.

[a] See also Chapter 1, *Protocol 3*, Chapter 4, *Protocol 2*.

2.2 Embedding YACs in agarose

This method is similar to that described for mammalian cells in *Protocol 1* except that the yeast cell wall must be digested prior to cell lysis.

Protocol 2. Embedding YAC DNA in agarose blocks[a]

Reagents

- AHC: 1.7 g Difco yeast nitrogen base (without amino acids, without $(NH_4)_2SO_4$, 6 g $(NH_4)_2SO_4$, 10 g Difco bacto casamino acids, 100 mg adenine sulfate, 50 mg tyrosine pH 5.8/litre
- SCE: 1 M sorbitol, 0.1 M sodium citrate, 0.06 M EDTA pH 7.0
- Yeast lytic enzyme: recombinant enzyme (ICN), zymolyase (Miles or Kirin Breweries), or non-recombinant lytic enzyme (ICN or Sigma)
- 0.45 M EDTA pH 9.0, 10 mM Tris–HCl pH 8.0, 7.5% (v/v) β-mercaptoethanol

Method

1. A 100 ml culture of a YAC-carrying yeast strain is grown to late log phase (10^8 cells/ml) at 30°C in AHC (ura−, trp−) media + 2% (w/v) glucose.
2. Pellet the cells by centrifugation at 4°C (1000 *g*, 10 min) and wash twice in 0.05 M EDTA pH 7.5.
3. Resuspend the cells in 3 ml 0.05 M EDTA and mix with 5 ml molten 1% (w/v) LMP ultrapure agarose, prepared in 0.125 M EDTA pH 7.5 and cooled to 42°C.
4. Add 1 ml SCE, 50 μl β-mercaptoethanol, and 60 U recombinant yeast lytic enzyme/zymolyase, or 100 U non-recombinant enzyme, to digest the yeast cell wall.
5. Aliquot into moulds and allow to set at room temperature.
6. Push the plugs into 0.45 M EDTA pH 9.0, 10 mM Tris–HCl pH 8.0, 7.5% (v/v) β-mercaptoethanol.[b] Incubate at 37°C overnight to form spheroplasts.

Protocol 2. *Continued*

7. Transfer plugs to NDS containing 1 mg/ml proteinase K. Incubate overnight at 50°C.
8. Store plugs in fresh NDS at 4°C.

[a] See also Chapter 4, *Protocol 8*.
[b] Take care as this is a lot of β-mercaptoethanol! Perform this and subsequent stages in a fume cupboard.

3. Resolving large DNA molecules by pulsed field gel electrophoresis (PFGE)

The resolution limit of normal agarose electrophoresis is reached when DNA molecules become too large to be sieved by the gel matrix (in practice > 50–100 kb). Schwartz and Cantor (6) described a form of agarose electrophoresis which can resolve DNA molecules of up to several megabases in size. The underlying principle of this form of electrophoresis is that DNA molecules are subject to obtuse or orthogonal electric fields so that the DNA molecules are forced to reorientate, within the agarose gel matrix, in response to the changing electric fields. Small DNA molecules are able to reorientate faster than larger ones (7). There are now many commercial PFGE systems available (8) which differ in the exact approach that they use to subject the DNA to the changing electric field, from physically moving the gel versus the electrodes, moving the electrodes versus the gel, and using hexagonal arrays of electrodes (contour-clamped homogeneous field electrophoresis—CHEF). This latter approach is the basis for one of the most popular commercially available PFGE systems—the Bio-rad CHEF-DRII. The specific running conditions described here apply to this piece of apparatus. For most purposes, brands of agarose used for normal gel electrophoresis can be used for casting a PFGE gel, however there are several specialized agarose products available which claim to facilitate the resolution of large DNA molecules. In general, low endosmosis agarose increases the speed of separation of large DNA molecules. However it should be noted that apparent fragment size, as judged by relative mobility, can vary depending on the type of agarose being used. Therefore, when constructing long-range restriction maps it is best to be consistent in the type of agarose used for running the different gels (9).

The sizes of the large DNA fragments generated by restriction enzyme digestion with the types of enzyme described in section 5 can vary by several orders of magnitude (tens to thousands of kilobases). Hence, running conditions for PFGE may be altered to optimize the resolution of DNA fragments within certain size ranges. Parameters that are routinely varied include the field strength, pulse time (including ramped pulse times), and run length. *Table 1* lists some suggested running conditions for resolving DNA fragments

Table 1. Suggested running conditions for PFGE of different sized DNA fragments

Size range of DNA fragments (kb)	Per cent agarose	Field strength V/cm	Initial pulse time (sec)	Final pulse time (sec)	Run time (h)
50–500	1	4.5	30	90	25
100–1000	1	4.5	100	150	42
300–2000	1	4.5	150	250	45
500–6000	0.6	1.5	1800	3600	140
50–6000	0.6	4.5 for	30	120	15
		then 1.5 for	120	3600	25

of varying size ranges using a CHEF apparatus. All gels are generally run at a temperature of 12–14°C by circulating the electrophoresis buffer (0.5 × TAE) through a cooler.

Appropriate size markers should be selected for the range of fragment sizes being resolved. These include oligomers of the lambda genome (50 kb upwards), chromosomes of the budding yeast *Saccharomyces cerevisiae* (90–2500 kb), and chromosomes of the fission yeast *Schizosaccharomyces pombe* (3000–6000 kb) (10). In addition several strains of *S. pombe* have been described that carry additional small minichromosomes that can be useful as size markers for PFGE. The preparation of *S. cerevisiae* chromosomes is essentially the same method as described for YACs in *Protocol 2*, except that a non-selective yeast medium such as YPD (1% (w/v) yeast extract, 2% (w/v) bacto peptone, 2% (w/v) glucose) is used in place of AHC. It is best to use a strain with a well defined electrophoretic karyotype such as YPH148 (11). The preparation of lambda oligomers from λ virions and *S. pombe* chromosomes are described in *Protocols 3* and *4*, respectively.

Protocol 3. Preparing oligomers of lambda phage DNA embedded in agarose

Equipment and reagents

- *E. coli* λ lysogen with a temperature-sensitive cI repressor e.g. λcI857Sam7
- L broth and L agar plates (12)
- TE: 10 mM Tris–HCl pH 8.0, 1 mM EDTA
- RNase A (10 mg/ml)
- DNase I (10 mg/ml)

Method

1. Streak out the lysogenic *E. coli* strain on LB agar plates at 32°C and 45°C to check for lack of growth at 45°C (i.e. that the cI repressor is inactivated at this temperature).

Protocol 3. *Continued*

2. Inoculate a single colony from the 32°C plate into 100 ml L broth, grow at 32°C with aeration to mid-log phase (OD 580 nm = 0.3).
3. Place the flask in a 45°C water-bath, or in a sink or bucket filled with water at slightly over 45°C. Place a thermometer into the culture. Swirl the flask in the water until the temperature of the culture reaches 45°C. Keep on swirling in the bath, maintaining the temperature of the culture, for a further 15 min.
4. Transfer the flask to a 39°C incubator and grow the culture for a further 2.5 h.
5. Pellet the cells at 4°C for 10 min at 10 000 *g* and resuspend in 8.8 ml TE.
6. Add 0.5 ml chloroform and shake at 37°C for 15 min to release the λ virions.
7. Add 20 μl of RNase A and DNase I (both at 10 mg/ml) and incubate for a further 15 min.
8. Pellet the debris at 4°C, 15 min at 12 000 *g*.
9. To 4.5 ml of supernatant add 4.5 ml of TE and 9 ml of 1% (w/v) LMP agarose cooled to 42°C.
10. Aliquot into plug moulds and allow to set.
11. Incubate plugs in NDS, 0.5 mg/ml proteinase K at 50°C for 48 h.
12. Store at 4°C in NDS.

Protocol 4. Preparing chromosomes of *S. pombe* in agarose plugs

Reagents

- SP1:1.2 M sorbitol, 50 mM sodium citrate, 50 mM sodium phosphate, 40 mM EDTA, pH 5.6
- Zymolyase

Method

1. Grow a 100 ml *S. pombe* culture at 32°C to 2 × 10^7 cells/ml. Harvest by centrifugation at 1000 *g*, for 10 min at 4°C.
2. Wash the cells three times in 50 mM EDTA pH 8.0.
3. Resuspend the cells in 1 ml SP1 and add 50 U of zymolyase and incubate at 37°C for 2 h.
4. Add 1 ml of molten 1% (w/v) LMP agarose, made up in 0.125 mM EDTA, and cooled to 42°C.

5. Aliquot into plug moulds and leave to set.
6. Incubate plugs in NDS containing 0.5 mg/ml proteinase K, at 50°C overnight.
7. Incubate in fresh NDS/proteinase K for a further 48 h.
8. Store plugs at 4°C in fresh NDS.

After electrophoresis is complete, the gel is stained with ethidium bromide and photographed. It can then be transferred to hybridization membrane by Southern transfer. It is preferable to use a reusable nylon membrane so that the blot can be hybridized sequentially with multiple probes to generate a restriction map. Transfer is essentially the same as for standard agarose gels except that partial depurination of the DNA in the gel, in 0.25 M HCl for 15 min, prior to denaturation assists in the transfer of larger DNA molecules.

4. Restricting DNA embedded in agarose

Constructing long-range restriction maps requires a battery of restriction enzymes that will produce large fragments. Many rare cutting restriction enzymes are available from different suppliers, including New England Biolabs (NEB), Boehringer Mannheim (BCL), Amersham International, and Stratagene. Restriction digests of DNA embedded in agarose are carried out as described in *Protocol 5* in buffers of the composition stipulated in the manufacturers' specifications and are at the recommended incubation temperatures for each enzyme.

Protocol 5. Digesting DNA embedded in agarose plugs[a]

Reagents

- Restriction buffer wash: 10 mM Tris–HCl pH 7.5, 10 mM $MgCl_2$, and either 0 mM, 50 mM, or 100 mM NaCl as appropriate for the enzyme being used
- Phenylmethylsulfonyl fluoride (PMSF): 20 mg/ml made up fresh in propan-2-ol (Sigma)
- TE: 10 mM Tris–HCl pH 8.0, 1 mM EDTA

Method

1. For each track incubate one-half of an agarose block in 5 ml TE containing 5 μl fresh PMSF (20 mg/ml in propan-2-ol).[b] Leave at 50°C for 30 min.
2. Repeat step 1, with fresh TE and PMSF.
3. Incubate each plug in 5 ml TE at 4°C for 30 min.
4. Repeat step 3.
5. Incubate in 1 ml of appropriate restriction buffer wash at 4°C for 30 min.

Protocol 5. *Continued*

6. Repeat step 5.
7. Add 100 μl of the restriction buffer recommended by the enzyme supplier. Add 10–20 U of enzyme if complete digestion is required and incubate between 4 h and overnight at the appropriate temperature. For partial digestion, both the amount of enzyme added and the time of incubation can be reduced.

[a] See also Chapter 1, *Procotol 5*.
[b] PMSF should be handled with care. It is rapidly inactivated in alkaline aqueous solution and can then be safely disposed of.

Problems can arise with restriction endonuclease cleavage of DNA embedded in agarose. If this happens an additional overnight incubation of the plugs with proteinase K in NDS may help. Additionally, with YAC DNAs there may have been inadequate digestion of the yeast cell wall prior to proteinasing. If problems persist, the PMSF may be too old and a fresh batch should be purchased. Addition of Triton X-100 to 0.01% (v/v) has been shown to aid digestion by *Not*I. We routinely add 5 mM spermidine–HCl to all digests in buffers with a NaCl concentration of greater than 50 mM.

DNA embedded in agarose can be digested with multiple enzymes sequentially. If the recommended restriction buffers of the enzymes chosen are incompatible, digest with the enzyme cutting in the lowest concentration of NaCl first, then wash the plug twice in 1 ml of the restriction buffer wash appropriate for the next enzyme to be used, before adding that enzyme and its appropriate incubation buffer and proceeding with the next digestion.

5. Choosing the right enzyme for the job

5.1 Generating large restriction fragments

The most obvious way to generate large restriction fragments for PFGE analysis is to use enzymes with large recognition sequences. The most extreme examples of these are the homing endonucleases encoded by the group I mobile introns found in some fungal nuclear and mitochondrial genomes, T4 phage, and the chloroplast and mitochondrial genomes of *Chlamydomonas*. The recognition sequences for these enzymes are long (10–19 bp) and unlike those for type II restriction endonucleases they are not necessarily symmetrical (13). Since there are few or no sites for these enzymes in many complex genomes, it has been proposed that their recognition sites could be inserted into the genome or into YACs to create targeted and unique restriction sites for map making (14). The yeast HO nuclease could be used in a similar way. This approach is akin is to that used by Chikashige *et al.* (15), who integrated *Not*I sites into the centromeres of *S. pombe* in order to map them. Recently, a sequence-specific cleavage method, utilizing the *E. coli* recA protein has been used to map several regions of the human genome (16). However, these are likely to remain specialized approaches for creating cleavage maps.

For type II restriction endonucleases, the largest recognition sites commonly available are 8 bp in length. These include those for: *Asc*I (GGCGCGCC), *Not*I (GCGGCCGC), *Fse*I (GGCCGGCC), *Srf*I (GCCCGGGC), *Pac*I (TTAATTAA), *Swa*I (ATTTAAAT), *Pme*I (GTTTAAAC), *Sce*83871 (CCTGCAGG), *Sgr*A1 (CA/GCCGGT/CG), and *Sfi*l (GGCCNNNNN-GGCC) (17). The uses of some of these enzymes are described in more detail below. Base modifications such as DNA methylation can also block cleavage by restriction enzymes, and indeed most of the CpG containing enzymes described below are inhibited by CpG methylation. This occurs naturally in some genomes, but can also be introduced enzymatically *in vitro* to generate larger restriction fragments (18)—see section 11.

5.2 CpG islands

In vertebrate DNA the dinucleotide CpG is present at only 25% of the expected frequency and most of the remaining genomic CpGs are methylated (19). Since many restriction enzymes are also sensitive to cytosine methylation, enzymes with sites containing one or more CpGs cut infrequently in vertebrate DNAs, even though those recognition sites may only be 6 bp long. Unmethylated CpG residues cluster in short 'CpG islands,' that are on average 1.5 kb long and are found at the 5′ end of many genes (20). In mammals, but not all vertebrates, CpG islands are also generally GC-rich (21). All of these properties combine to mean that CpG islands are the principal sites of cleavage for methylation-sensitive restriction enzymes with G + C-rich recognition sequences, containing one or more CpG residues (22). Hence, these enzymes can be used to create long-range restriction maps, which have the added bonus of highlighting the position of genes (approximately one-half of all mammalian genes, including all 'housekeeping genes' have islands). Conversely, once a CpG island is found on a map it is very likely to mark a gene, and is a widely used method for identifying genes in genomic DNA (23, 24).

The tendency of a CpG-containing restriction enzyme to cleave at islands depends on the total G + C composition of the recognition site and the number of CpG dinucleotides it contains, as well as its length. Enzymes can therefore be grouped according to these properties, with the different groups calculated, and indeed, observed to have different cutting characteristics and uses in mapping mammalian DNA (22). For example, the proportion of the genomic sites for an enzyme that are within CpG islands as opposed to in intergenic DNA determines how effectively the enzyme discriminates between island and non-island DNA. The average number of sites for a particular enzyme per island will determine the typical size of restriction fragment obtained, and how likely it is that the map that it generates will detect all the islands contained within it. The estimates of these properties, described here, are based upon the assumptions about the base composition of bulk human genomic DNA, and of the island fraction, used previously (20), an estimated average island size of 1.5 kb and a revised estimate of 45 000 for the total number of CpG islands in the human genome (25). To maximize the biologically relevant information that can be gleaned from a long-range restriction map of mammalian genomic DNA thought should be given to both the choice of enzymes, the source of the DNA, and the way in which the data is interpreted. In the following sections I will describe the uses of some groups of enzymes, in mapping the mammalian genome. For each group, the most common enzyme is indicated first, followed by its isoschizomers, and the recognition sequence.

5.2.1 *Asc*I (GGCGCGCC) and *Not*I (GCGGCCGC)

These enzymes have 100% G + C recognition sequence and two CpGs. As a consequence >90% of the sites for these enzymes in mammalian DNAs are in CpG islands. Hence the presence of a site for one of this pair of enzymes on a long-range map is a strong hint that there is an island there, and hence a gene. However, because islands are small and the sites for these enzymes are large (8 bp) not every island (only one in three on average) contains a site. This is in fact a higher frequency of sites per island than would be expected (22). This is illustrated in *Figure 1*, that shows the long-range restriction map of a part of human chromosome 17. There are probably nine (numbered) CpG islands on this map but only four of them contain an *Asc*I site, and similarly only four a *Not*I site.

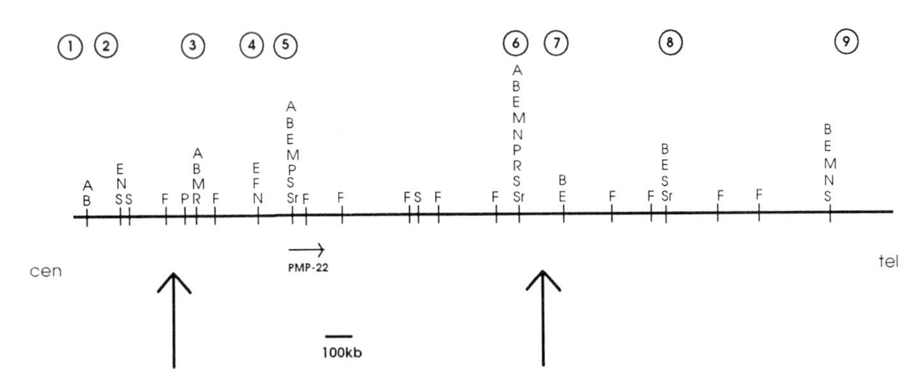

Figure 1. The long-range restriction map of the region of human 17q11.2 that contains the PMP-22 gene is shown. The data is drawn from the maps described in refs 34 and 47. A, *Asc*I; B, *Bss*HII; E, *Eag*I; F, *Sfi*I; M, *Mlu*I; N, *Not*I; P, *Pvu*I; R, *Nru*I; S, *Sac*II; Sr, *Srf*I. The position and direction of transcription of the PMP-22 gene is shown as are the sites (numbered and circled) of the nine potential CpG islands on this map. The 5′ end of the PMP-22 gene appears to be associated with island number 5. The vertical arrows indicate the proximal and distal boundaries of the 1.5 Mb region that is duplicated in CMT 1A (34, 47) and deleted in HNPP (35). The scale bar represents 100 kb.

Fragments generated by *Not*I and *Asc*I are, on the whole large (however see section 6), making these enzymes particularly useful for surveying large tracts of DNA and for detecting distant changes in a map, that signal, for example, the presence of a chromosome breakpoint (see section 8). It should also be noted that the internal six bases of the *Asc*I and *Not*I sites are *Bss*HII and *Eag*I sites, respectively (see below).

5.2.2. *Fse*I (GGCCGGCC) and *Srf*I (GCCCGGGC)

Like *Asc*I and *Not*I these enzymes have 8 bp recognition sequences that are 100% G + C. However they contain only one CpG dinucleotide. This means that, whilst they are calculated to have the same number of cutting sites per island as *Not*I and *Asc*I, 30% of the recognition sites are located in inter-island DNA. Only three out of the nine islands in *Figure 1* contain an *Srf*I site.

5.2.3 *Sfi*I (GGCCNNNNNGGCC)

Note that *Sfi*I is used at 50°C, and that although, like the previous two groups of enzymes, it has a 100% GC base composition in the 8 bp invariant part of its recognition sequence this does not include any CpG residues. This means that most *Sfi*I sites are not island-associated and that less than one in ten islands will contain an *Sfi*I site. This is backed-up by the map in *Figure 1*. In contrast to *Asc*I, *Not*I, and *Srf*I, there are 12 *Sfi*I sites on the map and only one of those is probably CpG island-associated. This illustrates the complementary distribution of enzyme sites that can be obtained using enzymes with different characteristics. This can be very useful in generating maps. For example, island number six of *Figure 1* contains the cutting sites for all of the rare cutting enzymes used to make this map, apart from *Sfi*I. Therefore it would have been hard to extend the map past this island were it not for the *Sfi*I fragment that spans both island numbers 6 and 7.

5.2.4 *Cpo*I/ *Csp*I/ *Rsr*II (CGGA/TCCG)

These isoschizomers recognize a 7 bp site with degeneracy at the central position. The site contains two CpG dinucleotides which results in a high percentage (73%) of sites calculated to be in CpG islands. Almost one-half of islands are expected to contain one of these sites. Note that incubation with *Csp*I is carried out at 30°C.

5.2.5 *Bss*HII (GCGCGC), *Eag*I/ *Ecl*XI/ *Bst*ZI/ *Eco*521/ *Xma*III (CGGCCG), and *Sac*II/ *Sst*II/ *Kpn*378I/ *Ksp*I (CCGCGG)

This group of enzymes is one of the most useful for long-range mapping and particularly for identifying the position of CpG islands in genomic DNA. Since, like *Asc*I and *Not*I, they have sites that are 100% G + C and contain two CpG dinucleotides, their cutting sites are predominantly (80%) within islands. However, as these sites are only 6 bp long they are more frequent, generating smaller fragments than *Not*I or *Asc*I with, on average, each island having one site for each of these enzymes. Coincident sites for these enzymes on long-range maps signal the presence of CpG islands and will identify most of the islands in any tract of the genome. In *Figure 1*, seven out of the nine potential islands contain a *Bss*HII site and seven an *Eag*I site. Only five of the islands contain a *Sac*II site and there a couple of extraneous *Sac*II sites as well.

Double digestion with combinations of these enzymes can confirm that the cutting sites seen are indeed clustered rather than staggered. It should be remembered that the limited level of resolution of PFGE means that what may appear to be a single cluster of sites may be two (or more) closely spaced islands (within a few kilobases of each other). Note that *Sac*II has some site preferences and is inhibited by NaCl concentrations greater than 10 mM and that digests with *Bss*HII are carried out at 50°C.

5.2.6 *Nae*I/ *Ngo*MI (GCCGGC), *Nar*I/ *Bbe*I/ *Ehe*I/ *Kas*I (GGCGCC), and *Sma*I/ *Cfr*9I/ *Psp*AI/ *Xma*I (CCCGGG)

Like the previous group of enzymes, the 6 bp recognition sites for *Nae*I, *Nar*I, and *Sma*I are 100% G + C and hence occur, on average once per island. However, these sites have only one CpG dinucleotide and this means that 50% of them are outside islands, although most of these sites will be methylated in genomic DNA. However, in cloned DNA where there is no CpG methylation, sites for these enzymes are just as likely to be found in inter-island DNA as in islands themselves. *Nae*I, and *Nar*I show site preferences.

5.2.7 *Mlu*I (ACGCGT), *Nru*I/ *Spo*I (TCGCGA), *Pvu*I/ *Bsp*C1/ *Xar*II (CGATCG), and *Spl*I (CGTACG)

The high G + C content of human CpG islands means that as As and Ts appear in the recognition sequence of an enzyme the proportion of sites outside islands increases. Despite the presence of two CpG residues in the recognition sequences of the enzymes listed here most sites are estimated to be in inter-island DNA and less than half of all islands will have a site for one of these enzymes. In fact, the observed proportion of sites for these enzymes in islands is even lower than expected (22). Sites for these enzymes in cloned DNA are not reliable pointers to the position of CpG islands. Additionally, because so many genomic sites are in inter-island DNA, and subject to variable methylation, these enzymes are not reliable for the detection of altered restriction fragments resulting from chromosome breakpoints. However, the variable methylation can be useful for establishing physical linkage.

5.2.8 Six cutter enzymes with 66.7% G + C but only one CpG

There are a large group of available enzymes that fall within this category, commonly used examples are *Sal*I (GTCGAC) and *Xho*I (CTCGACG). Their properties and uses are rather similar to those of the group of enzymes described above. Over 80% of the sites for enzymes in this category will be in inter-island DNA.

6. Genome organization

It has been estimated that there are approximately 45 000 CpG islands in the haploid mammalian genome (24) and hence their average spacing is calculated to be about one per 70 kb of DNA. However, analysis of different regions of the human and mouse genomes has shown that CpG islands are not evenly distributed. In some parts of the genome they occur frequently (every few tens or hundreds of kilobases) whilst elsewhere they are rare (< one per megabase). Regions of very frequent CpG islands include, the terminal regions of the short arms of human chromosomes 4 (the location of the Huntington's gene) and 16 (the location of the α-globin complex and autosomal dominant polycystic kidney disease), as well as the MHC complex at 6p21.3 (26–28). By contrast the DNA around the dystrophin gene contains very few CpG islands over a large stretch of genomic DNA (29). In general the CpG-rich compartments of the genome correspond to the cytogenetic R bands and the CpG-poor to the G bands (30). Therefore, prior knowledge of the chromosomal environment of a particular DNA marker may indicate the type of long-range map that it will generate.

Creating restriction maps in these different types of genomic region throws up different problems. In the CpG-poor regions very large restriction fragments are generated with many different enzymes and therefore large stretches of DNA can be analysed but the level of detail in them is low. Constructing long-range maps and detecting chromosome breakpoints in these regions of the genome can be done with few DNA markers as each marker can be used to probe relatively large stretches of DNA (29, 31). Mapping large stretches of mammalian genomic DNA in a CpG island-rich area of the genome can be a more labour intensive procedure since restriction fragments will be on the whole small, and sites may cluster. Consequently, a high density of DNA markers is usually needed for making these types of map, certainly higher than would be needed to make a map of the same size in an island-poor region of the genome. As well as being replete with CpG islands, the R band regions of the genome are also quite GC-rich (32). Hence the enzymes with 8 bp 100% A + T recognition sites (*Pac*I and *Swa*I) can be used to produce large restriction fragments in these genomic regions.

7. Physically linking up probes

The smallest restriction fragment to which two DNA probes can co-hybridize gives a maximum value for the amount of DNA which must separate them. Two probes may appear to hybridize to the same restriction fragment but in fact be located on two different fragments, of sizes that are so similar that they cannot be distinguished by PFGE. Independent corroboration of physical linkage is therefore desirable, such as co-hybridization of two probes to restriction fragments generated by more than one enzyme (particularly those with complementary properties) or to a ladder of partial digest fragments (see section 9), or independent evidence that indicates the close association of two probes, such as close genetic linkage or presence of both markers on the same YAC clone. If two probes are thought to be physically close to one another, but separated by a CpG island that contains many rare cutting restriction sites, it may be useful to try an enzyme with more of a propensity for cutting outside of islands (such as *Sfi*I or *Mlu*I) in order to find a restriction fragment that will link the two probes (*Figure 1*). Specialized linking probes that span across rare cutting enzyme sites are especially useful in the construction of long-range maps and in joining up adjacent fragments generated by rare cutter enzymes (33).

CpG-containing restriction sites outside islands, are subject to variable degrees of methylation, both between individuals and between different tissues of the same individual. Thus hybridization to, for example, an *Mlu*I digest of DNAs from different sources will often produce a complex series of bands for any one probe (22). Another probe, truly linked to the first, will produce the same or a closely related fingerprint of bands. This is a very good way of confirming physical linkage of two probes using a single restriction enzyme.

8. Chromosome rearrangements

In many cases disease loci have been identified by the segregation of a mutant phenotype with a cytologically visible chromosome abnormality such as a deletion or translocation. The smallest restriction fragment that can identify a chromosome breakpoint indicates the maximum distance between the probe used to detect the fragment and the breakpoint itself. Enzymes with no CpG dinucleotides in their recognition sites or those which are predominantly island associated are recommended for this job so that you can be confident that any novel fragments detected in this exercise are really due to a restriction fragment being disrupted by a chromosome rearrangement rather than due to variable methylation of a non-island associated CpG dinucleotide in the cutting site of the enzyme.

9. Partial digestion

Often the efficiency of digestion of DNA in agarose is not optimal and partial digestion products are observed even when all enzyme sites are capable of being cut. Partial digestion does in fact have advantages in the construction of long-range maps. It enables probes on adjacent restriction fragments to be linked by common hybridization to a partial restriction fragment, and hence extends long-range maps over longer distances.

A good example of the use of partial digestion has been in delimiting the chromosome rearrangements in Charcot-Marie-Tooth disease type 1A (CMT 1A). This is the most common inherited human peripheral neuropathy and has been found to be associated with a 1.5 Mb duplication of a region of human chromosome 17p11.2 that includes the gene encoding the peripheral myelin protein-22. The long-range restriction map of the CMT 1A region is shown in *Figure 1*. Delineating these duplications depended on the use of partial digests since the region contains several CpG islands that each contain many of the cutting sites for CpG island-associated enzymes. Consequently, using the PMP gene as a probe on *Asc*I or *Not*I digests, for example, does not allow detection of the limits of the duplication unit (indicated by the vertical arrows in *Figure 1*) either 5′ or 3′ to the gene, since there are intervening CpG islands. Rather partial digests with a variety of rare cutting enzymes were used (34). Subsequently, repeat sequences were identified flanking the 1.5 Mb monomer of the duplication, and it was postulated that these repeats direct misalignment of homologues and consequent unequal crossing-over of this region to give either the duplication of CMT 1A or the reciprocal product (a deletion of the same region) found in another autosomal dominant neuropathy HNPP (35).

Partial digests may occur naturally, especially for enzymes such as *Sal*I and *Mlu*I where variable methylation of sites may occur within a population of cells. Alternatively, partial digestion may be achieved by limiting the amount of enzyme during the digestion or by enzymatic methylation of sites (18).

10. Indirect end-labelling

Indirect end-labelling of partially restricted DNA is a convenient way of mapping linear DNA molecules (36). For DNA cloned in YACs this can readily be done by hybridizing partial digests of YAC DNA with probes derived from either the *trp* or *ura* YAC vector arms.

For indirect end-labelling of genomic DNA, use can be made of the ends of chromosomes (telomeres). These may either be in the form of the naturally occurring telomere or telomeres that have been artificially introduced into chromosomes on fragmentation vectors (37). An example of the use of natural telomeres for mapping by indirect end-labelling is a study of the pseudoautosomal regions of the human sex chromosome using probes for the telomeres of the short arms of the X and Y chromosomes. Partial digestion allowed maps of over 2 Mb to be built up for each chromosome, encompassing the entire pseudoautosomal region of the X and Y and in fact crossing the pseudoautosomal boundary into X and Y-specific DNA (38).

Using chromosome telomeres as the end-points for indirect end-labelling limits the genomic regions that can be mapped in this way. A more general approach is to use a restriction site to generate the end-point for the map. For example, complete digestion of the DNA with an enzyme such as *Not*I could be followed by partial digestion with a complementary type of enzyme such as *Sfi*I. A *Not*I end clone would then be the ideal hybridization probe for generating a unidirectional long-range map from the *Not*I site.

11. Methylation

The previous sections have already highlighted the problems of interpretation that can arise from variable methylation of cutting sites for CpG-containing enzymes (usually those that are not in islands). Methylation can also be useful in establishing physical linkage between probes and in extending long-range maps past islands. The methylation state of the DNA can be altered experimentally, by enzymatic methylation of the DNA *in vitro* (18) or by demethylation *in vivo*, by culturing cells in the presence of 5-azacytosine (39).

Most CpG islands appear to remain unmethylated in all tissues of the animal regardless of the state of expression of their associated genes. One prominent exception to this is in female mammals where islands on the inactive X chromosome become methylated (40). Methylation of some CpG islands is also seen in culture, and the extent of this tends to reflect the length of time for which the cell line has been established (41).

Apart from the inactive X, there are other documented instances of CpG island methylation in somatic tissue. In several Wilm's tumours and retinoblastomas, methylation of the CpG islands associated with the respective tumour-predisposing genes has been seen and proposed to be a contributory step along the road to tumour formation (5, 42). One particularly fascinating example of CpG island methylation is in the Fragile X syndrome, where expansion of trinucleotide repeats at the 5′ end of the FMR-1 gene is accompanied by methylation of the associated CpG island (43).

The biological source of DNA is therefore an important consideration in the analysis of vertebrate DNA. To detect all the CpG islands in a particular region it is best to use tissue direct from the animal or from a recently established cell line.

Analysis of YACs precludes any blockage of enzyme cutting by CpG methylation and so many more recognition sites for some enzymes are revealed in maps generated from YAC DNA rather than from genomic DNA. These are generally isolated sites for enzymes, such as *Mlu*I, which have many non-CpG island recognition sites, but there may be some clusters of sites for enzymes such as *Bss*HII, *Eag*I, or *Sac*II that are methylated in some or all tissues of the animal and are only cleaved in cloned DNA (44). The biological significance of these sites is not known, they may represent islands associated with recently inactivated pseudogenes or just be fortuitous clustering of sites in a G + C-rich region of DNA.

12. Mapping in non-vertebrate genomes

The discussion of the cutting properties of CpG-containing restriction enzymes has focused on the genomes of vertebrates, which have wide-spread CpG methylation and islands of unmethylated DNA. The cutting characteristics of these enzymes are also affected by the base composition of the islands. Figures discussed in section 5 are based on the properties of human CpG islands, which are quite GC rich. CpG islands of other vertebrates may have quite different base compositions (21). It should also be borne in mind that long-range restriction maps of regions of synteny may not be directly comparable between vertebrates since, for example, CpG islands seem to be being lost from the mouse genome at a faster rate than from the human genome. Consequently, approximately one in five of the genes that have an island in human have no island in the mouse (25). These tend to be genes with a tissue-specific pattern of expression.

Plant DNA has methylation not only at CpGs but also at CpXpGs. This means that enzymes such as *Not*I (which contains both two CpGs and two CpXpGs in its site) should still generate large restriction fragments in plant DNAs. *Asc*I, which has the same cutting properties in mammalian DNA as *Not*I has no CpXpGs. Indeed *Not*I does seem to generate large restriction fragments in most plant species (45). There do seem to be exceptions however. The rice genome, for example, is cut quite frequently by all the CpG and CpXpG-containing enzymes tested, making construction of long-range maps in this organism difficult (46).

Acknowledgement

W. A. B. is a Lister Institute Research Fellow.

References

1. Koob, M. and Szybalski, W. (1992). In *Methods in enzymology* (ed. R. Wu), Vol. 216, pp. 13–19. Academic Press, London.
2. Pasero, P., Sjakste, N., Blettry, C., Got, C., and Marilley, M. (1993). *Nucleic Acids Res.*, **21**, 4703.
3. Guzman, P. and Ecker, J. R. (1988). *Nucleic Acids Res.*, **16**, 11091.
4. Guidet, F., Rogowsky, P., and Langridge, P. (1990). *Nucleic Acids Res.*, **18**, 4955.
5. Royer-Pokora, B., Ragg, S., HecklOstreicher, B., Held, M., Loos, U., Call, K., *et al.* (1991). *Genes Chromosomes Cancer*, **3**, 89.
6. Schwartz, D. C. and Cantor, C. R. (1984). *Cell*, **37**, 67.
7. Smith, S. B., Aldridge, P. K., and Callis, J. B. (1989). *Science*, **243**, 203.
8. Anand, R. and Southern, E. M. (1990). In *Gel electrophoresis of nucleic acids: a practical approach*, 2nd edn (ed. D. Rickwood and B. D. Hames), pp. 101–23. IRL Press, Oxford.
9. Upcroft, P. and Upcroft, J. A. (1993). *J. Chromatogr.*, **618**, 79.

10. Vollrath, D. and Davis, R. W. (1987). *Nucleic Acids Res.*, **15**, 7865.
11. Rimm, D. L., Pollard T. D., and Hieter, P. (1988). *Chromosoma*, **97**, 219.
12. Sambrook, J., Fritsch, E. F., and Maniatis, T. (ed.) (1989). In *Molecular cloning: a laboratory manual*, Vol. 3, pp. A1. Cold Spring Harbor Press, NY.
13. Marshall, P. and Lemieux, C. (1992). *Nucleic Acids Res.*, **20**, 6401.
14. Puchta, H., Dujon, B., and Hohn, B. (1993). *Nucleic Acids Res.*, **21**, 5034.
15. Chikashige, Y., Kinoshita, N., Nakaseko, Y., Matsumoto, T., Murakami, S., Niwa, O., *et al.* (1989). *Cell*, **57**, 739.
16. Ferrin, L. J. and Camerini-Otero, R. D. (1994). *Nature Genet.* **6**, 379.
17. Roberts, R. J. and Macelis, D. (1991). *Nucleic Acids Res.*, **19S**, 2077.
18. Hanish, J., Rebelsky, M., McClelland, M., and Westbrook, C. (1991). *Genomics*, **10**, 681.
19. Bird, A. P. (1986). *Nature*, **321**, 209.
20. Bird, A. P. (1987). *Trends Genet.*, **3**, 342.
21. Cross, S., Kovarik, P., Schmidtke, J., and Bird, A. (1991). *Nucleic Acids Res.*, **19**, 1469.
22. Bickmore, W. A. and Bird, A. P. (1992). In *Methods in enzymology* (ed. R. Wu), Vol. 216, pp. 224–44. Academic Press, London.
23. Abe, K., Wei, J.-F., Wei, F.-S., Hsu, Y.-C., Uehara, H., Artzt, K., *et al.* (1988). *EMBO J.*, **7**, 3441.
24. Sargent, C. A., Dunham, I., and Campbell, R. D. (1989). *EMBO J.*, **8**, 2305.
25. Antequera, F. and Bird, A. (1993). *Proc. Natl Acad. Sci. USA*, **90**, 11995.
26. Baxendale, S., MacDonald, M. E., Mott, R., Francis, F., Lin, C., Kirby, S. F., *et al.* (1993). *Nature Genet.*, **4**, 181.
27. Harris, P. C., Barton, N. J., Higgs, D. R., Reeders, S. T., and Wilkie, A. O. M. (1990). *Genomics*, **7**, 195.
28. Campbell, R. D. (1993). In *Genome analysis: regional physical mapping*, Vol. 5 (ed. K. E. Davies and S. M. Tilghman), pp. 1–34. Cold Spring Harbor Press, NY.
29. Burmeister, M., Monaco, A. P., Gillard, E. F., van Ommen, G.-J. B., Affara, N. A., Ferguson-Smith, M. A., *et al.* (1988). *Genomics,* **2**, 189.
30. Craig, J. M. and Bickmore, W. A. (1994). *Nature Genet.*, **7**, 376.
31. Ward, J. R. T., Cottrell, S., Thomas, H. J. W., Jones, T. A., Howe, C. M., Hampton, G. M., *et al.* (1993). *Genomics,* **17**, 15.
32. Saccone, S., de Sario, A., della Valle, G., and Bernardi, G. (1992). *Proc. Natl Acad. Sci. USA,* **89**, 4913.
33. Frischauf, A.-M. (1989). *Technique,* **1**, 3.
34. Pentao, L., Wise, C. A., Chinault, A. C., Patel, P. I., and Lupski, J. R. (1992). *Nature Genet.,* **2**, 292.
35. Chance, P. F., Abbas, N., Lensch, M. W., Pentao, L., Roa, B. B., Patel, P. I., *et al.* (1994). *Hum. Mol. Genet.,* **3**, 223.
36. Smith, H. O. and Birnsteil, M. L. (1976). *Nucleic Acids Res.,* **3**, 2387.
37. Farr, C. J., Stevanovic, M., Thomson, E. J., Goodfellow, P. N., and Cooke, H. J. (1992). *Nature Genet.,* **2**, 275.
38. Brown, W. R. A. (1988). *EMBO J.,* **7**, 2377.
39. Dobkin, C., Ferrando, C., and Brown, W. T. (1987). *Nucleic Acids Res.,* **15**, 3183.
40. Keith, D. H., Singer-Sam, J., and Riggs, A. D. (1986). *Mol. Cell. Biol.,* **6**, 4122.
41. Antequera, F., Boyes, J., and Bird, P. (1990). *Cell,* **62**, 503.

42. Sakai, T., Toguchida, J., Ohtani, N., Yandell, D. W., Rapaport, J. M., and Dryja, T. P. (1991). *Am. J. Hum. Genet.*, **48**, 880.
43. Davies, K. (1991). *Nature,* **351**, 439.
44. Anand, R., Ogilvie, D. J., Butler, R., Riley, J. H., Finniear, R. S., Powell, S. J., *et al.* (1991). *Genomics,* **9**, 124.
45. Ganal, M. W., Bonierbale, M. W., Roder, M. S., Park, W. D., and Tanksley, S. D. (1991). *Mol. Gen. Genet.*, **225**, 501.
46. Wu, K.-S. and Tanksley, S. D. (1993). *Plant Mol. Biol.,* **23**, 243.
47. Timmerman, V., Nelis, E., Van Hul, W., Nieuwenhuijsen, B. W., Chen, K. L., Wang, S., *et al.* (1992). *Nature Genet.,* **1**, 171.

8

Genetic mapping with microsatellites

ISAM S. NAOM, CHRISTOPHER G. MATHEW, and MARGARET-MARY TOWN

1. Introduction

Simple tandemly repeated sequences, or microsatellites, are ubiquitous in the genomes of a wide range of organisms, and the number of repeats within many of them is highly variable in the population of a particular species. The advent of the polymerase chain reaction (1) provided the means for rapid analysis of the repeat number, and several groups showed that these sequences were likely to provide a rich source of very informative markers for genetic mapping (2–4).

The most common microsatellites in mammals are of the form $(dC\text{–}dA)_n$.$(dG\text{–}dT)_n$, which occur on average once in every 50 kb of the human genome. About 40% of cosmid clones will contain a CA repeat, and most YAC clones will contain several. Microsatellites with ten or fewer copies of the repeat are generally not polymorphic. However, those with 20 or more repeats are often highly polymorphic, with as many as 10–15 different alleles being detectable in the population. Consequently, a high proportion (70–90%) of the population will be found to be heterozygous when typed with such a marker; i.e. they will have two alleles of different repeat number. This degree of heterozygosity makes microsatellites much more powerful genetic markers than conventional restriction fragment length polymorphisms (RFLPs), which generally have only two alleles, and hence a maximum theoretical heterozygosity of 50%. Polymorphic repeats with tri-, tetra-, or pentanucleotide motifs also occur, but at a somewhat lower frequency than CA repeats. The size of each allele of the microsatellite is determined by amplification of the repeat with PCR primers from the unique sequences which flank it, followed by gel electrophoresis.

Microsatellite polymorphisms (MSPs) are now used for a wide range of applications in genetics, including the construction of genetic linkage maps, linkage mapping of disease genes and quantitative traits, diagnosis of genetic disorders, studies of loss of heterozygosity in tumours, paternity testing, forensic analysis, and selective breeding in the dairy and beef industries. This chapter will focus on strategies and protocols for efficient genetic linkage analysis by the simultaneous analysis of multiple MSPs.

2. Genetic mapping

2.1 Linkage analysis

Genetic linkage analysis is based on the principle that if two genes or DNA segments are located close to each other on the same chromosome, they are likely to be inherited together. The greater the physical distance between them, the more likely they are to be separated from each other by recombination during the formation of the gametes at meiosis. The distance between them on the genetic map is thus determined by the frequency with which they recombine, and is generally expressed either as the recombination fraction (θ) or as the percentage recombination in centimorgans (cM). 1 cM corresponds on average to a physical distance of 1 Mb in the human genome, but there are both recombination 'hotspots' and 'coldspots', where 1 cM covers much less or much more than 1 Mb respectively. Linkage between two genetic loci is established when they show significant cosegregation in offspring. The general test of significance is a LOD score of + 3.0, which corresponds to odds of 1000:1 supporting linkage, and a posterior probability of 95% that the two loci are linked. A detailed discussion of linkage theory has been published elsewhere in this series (5).

High-resolution genetic linkage maps composed entirely of MSPs have been constructed for all human chromosomes by the Genethon group in Paris (6). Their latest map is composed of over 2000 markers, with an average spacing of less than 3 cM. This map is an important landmark in human genetics since it provides both an efficient means of mapping genes responsible for human disease and a set of ordered genetic markers which can be used as a framework for the construction of a continuous physical map of all of the chromosomes. A very high-resolution linkage map of the mouse has been constructed which contains over 4000 MSPs and an average spacing of 0.35 cM (7). Microsatellite maps of the rat genome (8) and of the X chromosome of the mosquito (9) are also available.

2.2 Mapping disease genes

One of the most important applications for the microsatellite markers is to map genes which confer predisposition to disease in mammals. The following discussion will relate to mapping human disease genes, but significant progress has also been made in localizing such genes in mammals, including those associated with diabetes in the mouse (10), and hypertension in the rat (11).

The chromosomal localization of human single gene disorders by linkage analysis is now a relatively straightforward procedure, provided that DNA from a sufficient number of meioses are available in families affected with the disorder, that a secure diagnosis can be made in the affected individuals, and that the disorder is not genetically heterogeneous. If the disorder is caused by mutations at more than one genetic locus, linkage will be more difficult to detect since positive LOD scores at a linked locus in one family will be cancelled by negative scores in another family which is linked to a different locus. However, some linkages have been detected even where heterogeneity exists.

The strategies which can be adopted for linkage mapping of disease genes have been reviewed in detail (12, 13) and will be outlined briefly here. In the candidate gene approach, markers from within or around a gene which is considered likely to be involved in the disorder are tested for linkage to the clinical phenotype. In the absence of strong candidate genes, genome-wide searches are made with markers from all human autosomes. This search may be by a 'shotgun' approach, in which any available markers are tested irrespective of their chromosomal location. This method is inefficient since markers which are close to other previously typed markers already known not to be linked to the disease gene will be tested. The most efficient method is a systematic search of each chromosome with markers spaced at intervals of 15–20 cM. In this way, intervals between successive markers can be excluded from containing the disease gene. Updated information on all known human polymorphisms and their map location is available from the Genome Database (see Appendix for details).

The use of microsatellite polymorphisms for a genome-wide linkage search has several advantages. They are highly informative, so a relatively small number of families will be sufficient to prove or disprove linkage. They can be typed quickly by PCR, using only small amounts of DNA, and even using low molecular weight DNA isolated from archival material such as paraffin-embedded sections. Also, multiple MSPs can be analysed simultaneously, since several loci can be amplified in a single PCR and, provided that their allele sizes do not overlap, analysed on the same lane of a gel. These excellent properties of MSPs, and their ubiquity in the human genome, has encouraged the belief that it might now be possible to map genes which confer predisposition to common disorders such as heart disease, diabetes, and psychiatric illness. Since these are almost certainly multifactorial disorders, the analysis of much larger numbers of families will be required to identify the genes involved. For example, typing 200 sibling pairs with a common disorder with 200 markers will require the generation of 80 000 genotypes! Clearly the efficiency of the analysis must be maximized, and the process should be automated wherever possible.

3. Isolation and analysis of microsatellites

3.1 Isolation and characterization of MSPs

CA repeats can be isolated by screening grids of phage, cosmid, or YAC clones with a $(CA)_{15}$ oligonucleotide. Clones which show the strongest hybridization signal can then be selected for further analysis, since they will generally have the highest number of repeats. Detailed protocols have been described in a previous volume in this series (14). Briefly, CA-positive clones are digested with frequent cutter restriction enzymes and subcloned into plasmid vectors. CA-positive subclones are then sequenced to determine repeat number and the sequence of single copy DNA flanking the repeat. Those which contain ten or more repeats are then amplified in DNA samples from 10–20 unrelated individuals to establish whether they are polymorphic. Tedious sequencing of relatively large subclones to determine the sequences flanking the repeat can be avoided by using anchored sequencing primers of the form $(5'\text{dG–dT}3')_7\text{dX}$ (where X = A, G, or T), $(5'\text{dT–dG}3')_7\text{dX}$, $(5'\text{dC–dA}3')_7\text{dX}$, and $(5'\text{dA–dC}3')_7\text{dX}$ to sequence the subclones (15). Microsatellites containing tri-, tetra-, or pentanucleotide repeats can be screened for by use of a cocktail of oligonucleotides which contain the sequences commonly found in such repeats (16).

3.2 Options for analysis of MSPs

A variety of options exist for the analysis of MSPs. The simplest method is to electrophorese the PCR products on a 15–50 cm vertical, non-denaturing polyacrylamide gel, and detect the alleles by staining with ethidium bromide. This method is suitable for the analysis of MSPs with a repeat unit of more than two nucleotides, provided that the size of the product is less than about 200 bp. Alleles which differ by only one repeat unit will be difficult to resolve in larger PCR products. Some groups have been able to analyse CA repeats by this method (17), but we have found that non-denaturing gel electrophoresis of these markers produce additional conformers or heteroduplexes, which complicate the interpretation of the gel. The gels can also be stained with silver (17), which provides greater sensitivity of detection than ethidium bromide but also produces the more complex patterns associated with non-denaturing gels.

The approach most commonly used for the analysis of CA repeats is radioactive labelling of the PCR product, followed by electrophoresis on denaturing polyacrylamide gels. Two different methods are available. In the 'internal' labelling method, a single labelled deoxynucleoside triphosphate such as $[\alpha\text{-}^{32}\text{P}]$dCTP is added to the PCR. However, the $[\text{CA}]_n$ and the $[\text{TG}]_n$ strands will be separated by electrophoresis on denaturing gels, so that each allele will be detected as two bands. Further complexity is introduced into the gel pattern by the 'stutter' or shadow bands which are seen with these markers (14), which are one or two bases larger or smaller than the allele itself. These are believed to be caused by template-independent addition of one or more nucleotides to the 3′ end of the product, and slipped-strand mispairing during the PCR. We therefore favour the other option for radioactive analysis of these markers, which involves end-labelling one of the primers with [γ-^{32}P]ATP. This produces a much less complex gel pattern than the internal labelling method, and is consequently easier to interpret. Detailed protocols for the end-labelling method will be presented in section 4.

The most recent method developed for MSP analysis is the use of fluorescent dyes for the detection of alleles. The dyes offer greater sensitivity than other non-radioactive methods and are obviously less hazardous to use than radioactivity, but their main advantage is that PCR primers can be labelled with dyes which have different absorption spectra, thus permitting the simultaneous analysis of MSPs with alleles which overlap in size (18). The Applied Biosystems Automated DNA Fragment Analyser, for example, uses a red dye for an internal molecular weight marker, and three other colours for the MSP primers. This multicolour system, together with the improved resolution afforded by 'real time' capture of the data by a laser scanner, allows a much higher throughput of MSPs than is possible with the other methods of analysis. Experimental details for the use of this system are described in section 5.

Finally, some additional steps can be taken to increase the efficiency of MSP analysis. The PCR process allows amplification of multiple loci in a single reaction tube. We have combined up to eight CA repeat markers in a single multiplex PCR, but the upper limit may be much higher. The main constraints are that the primers should not have substantial regions of complementarity, and should direct specific amplification at the same annealing temperature. Multiplex PCR produces major savings of time, reagents, equipment utilization, and of sample DNA, but a balance must be struck between the effort expended on getting a set of different primers to PCR together, and the effort required to PCR them in separate reactions. A further increase in throughput can be achieved by automating the PCR process. PCRs can be set-up in 96-well microtitre plates using a robotic pipetting instrument such as the Biomek (Beckman Instruments), and amplified on thermal cyclers with a 96-well microtitre format.

4. Radioactive typing of MSPs using end-labelling

4.1 Selection of microsatellite markers

The selection of suitable markers is based on the following parameters:

(a) Heterozygosity. Markers with a heterozygosity of greater than 0.7 are preferred.

(b) Genetic distance between markers. This is determined by the linkage search strategy (see section 2).

   i. Markers for systematic linkage search are selected with a maximum spacing of 15–20 cM, so that the distance between the disease gene and linked markers would not on average be more than 10 cM (see *Figure 1* for an example of a set of markers used for a linkage search of chromosome 6p).

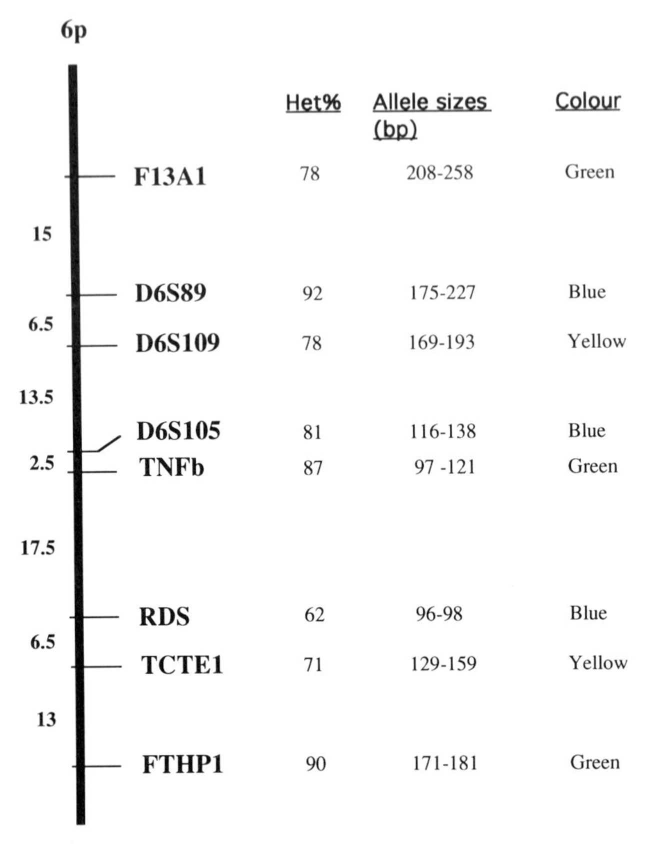

Figure 1. Sex averaged genetic map of the short arm of chromosome 6 demonstrating the choice of markers used in a search for a Crohn's disease gene. An additional marker, TNFb, is included in the HLA region as this was a candidate gene. Genetic distances between the microsatellite loci are given in centimorgans (cM). The heterozygosity, allele size range, and colour of the fluorescent label for each marker are shown to the right of the map.

ii. In the candidate gene approach, microsatellite markers within or flanking the candidate gene are selected, in order to maximize any potentially positive LOD scores.

(c) Allele size. Sets of markers with non-overlapping allele size ranges are selected (based on published values), since they can be electrophoresed simultaneously in a single gel lane. Leave a gap of 15 bases between the allele size ranges of each marker to allow for previously undetected alleles. A maximum of three or four markers can be analysed in one gel lane using the radionuclide labelling system.

4.2 Establishing the PCR conditions

Each marker is tested initially for specific amplification as described in *Protocol 1*. If three or four markers are to be run in a single gel lane, they should be tested to establish whether they can be amplified together in a single reaction. This is done on a trial and error basis, and by taking into account published annealing temperatures for the markers.

Protocol 1. Basic PCR method

Equipment and reagents

- Thermal cycler for DNA amplification
- T(0.1)E buffer: 1 mM Tris–HCl pH 8.0, 0.1 mM EDTA
- 5 mM dNTP mix: mix 25 μl of 100 mM stock solution of each dNTP and add 400 μl of T(0.1)E buffer
- 10 × *Taq* polymerase buffer: 670 mM Tris–HCl pH 8.8, 166 mM $(NH_4)_2SO_4$, 15 mM $MgCl_2$, 1.7 mg/ml nuclease-free bovine serum albumin (BSA)—add 3.5 μl of mercaptoethanol to 0.5 ml aliquots of the buffer
- PCR primers: working solution of 10 pmol/μl in T(0.1)E
- *Taq* DNA polymerase (5000 U/ml)
- Mineral oil (Sigma)
- Genomic DNA samples at a concentration of about 50 ng/μl
- 5% polyacrylamide or 2% agarose minigel

Method

1. Make up a PCR mixture for the required number of 25 μl reactions. Each PCR should contain:
   - 1 μl of each primer solution
   - 1 μl 5 mM dNTP mix
   - 2.5 μl 10 × *Taq* polymerase buffer
   - 14.2 μl T(0.1)E buffer (reduce the volume if more than one primer pair is being used)
   - 0.3 μl (1.5 U) *Taq* DNA polymerase (add last)

   Vortex gently and add 20 μl of the mixture to 5 μl of genomic DNA (250 ng). Overlay the mixture with 30 μl of mineral oil. Spin in a microcentrifuge for 15 sec. Vortex and spin again for 10 sec. Place the tubes in a thermal cycler.
2. Denature the DNA at 94°C for 5 min. Then perform 30 cycles of PCR using the following temperature profile:
   - denaturation 94°C, 45 sec
   - primer annealing[a] 55°C, 45 sec
   - primer extension 72°C, 2 min
   - a final extension cycle at 72°C for 5 min
3. Run 10 μl aliquots of the trial reactions on a non-denaturing 5% poly-

Protocol 1. *Continued*

acrylamide minigel or a 2% agarose minigel at 90 V for 1–2 h. Visualize the DNA fragments under UV light after staining with ethidium bromide.

[a] The annealing temperature will vary for different primers.

Since the multiplex PCR is a compromise of amplification conditions, variations occur in the quality of results for each pair of primers. Non-specific bands or multiple stutter bands can be seen on radioactive gels, which are not apparent on the trial agarose gels. These bands can make the identification of true alleles difficult, but it is possible to reduce the intensity of the non-specific bands and to optimize the multiplex PCR conditions by:

- reducing the primer concentration
- increasing the primer annealing temperature
- reducing PCR cycle number
- titrating the $MgCl_2$ concentration

In spite of optimizing these conditions to accommodate multiplex amplification, some marker loci will give consistently poor results and require individual amplification.

4.3 Labelling and amplification

5′ end-labelling with polynucleotide kinase and [γ-^{32}P]ATP is performed as described in *Protocol 2*.

Protocol 2. End-labelling with $\gamma-^{32}P$

Reagents

- See *Protocol 1*
- 10 × kinase buffer: 0.5 M Tris–HCl pH 7.6, 0.1 M $MgCl_2$, 50 mM dithiothreitol, 1 mM spermidine, 1 mM EDTA
- T4 polynucleotide kinase (10 000 U/ml)
- PCR forward primer: 50 pmol/μl in T(0.1)E (the reverse primer can be labelled instead, but one or the other should be labelled consistently)
- [$\gamma-^{32}$P]ATP (3 000 Ci/mmol)

Method

1. Set-up sufficient labelling reaction mix for the required amount of primer. 5 μl of this mix will be required for a single PCR.[a] Each labelling reaction should contain:
   - 0.1 μl primer
   - 0.25 μl T4 kinase buffer
   - 0.2 μl T4 polynucleotide kinase

- 0.1 μl [γ-^{32}P]ATP
- 1.85 μl T(0.1)E

Incubate the mixture at 37°C for 2–3 h, then add 2.5 μl of T(0.1)E buffer.

2. Prepare a PCR reaction pre-mix containing the reverse primer. This mix is the same as the PCR mix in *Protocol 1*, except that each reaction should contain 5 pmol of reverse primer and no forward primer.
3. Add 15 μl of reverse primer mix to 5 μl of genomic DNA. Then add 30 μl mineral oil, vortex gently, centrifuge briefly, and add the 5 μl of end-labelled primer.
4. Carry out PCR amplification as described in *Protocol 1*, but reduce the cycle number to 21–25 cycles (the cycle number should be adjusted to produce sufficient yield of PCR product with a minimum of additional bands).

[a] Labelled primers and PCR products can be stored at −20°C for up to three days before use.

4.4 Electrophoresis and autoradiography

The PCR products are now electrophoresed on a denaturing polyacrylamide gel to separate the alleles and to determine their size (*Protocol 3*). The alleles are detected by autoradiography (*Protocol 4*).

Protocol 3. Electrophoresis of DNA in polyacrylamide denaturing gels

Equipment and reagents

- Vertical gel electrophoresis apparatus (the standard apparatus used for DNA sequencing is adequate, but a 50 cm gel is preferred for multiplexes of several MSPs)
- 10 × TBE buffer: 0.89 M Tris, 0.89 M boric acid, 2 mM EDTA pH 8.0
- 6% denaturing acrylamide gel mix: 75 ml 40% acrylamide/bisacrylamide (19:1), 210 g urea, 25 ml 10 × TBE—make up to 500 ml with distilled water
- Formamide dye mix: 8 ml 80% (v/v) deionized formamide, 560 μl 10 × TBE, 200 μl 50 mM EDTA, 10 mg 0.1% (w/v) xylene cyanol, 10 mg 0.1% bromophenol blue—add distilled water to a final volume of 10 ml
- M13 sequencing ladder (16)
- TEMED (*N,N,N′,N′*-tetramethylethylenediamine)
- 10% ammonium persulfate

Method

1. Set-up glass plates with 0.4 mm spacers.
2. Mix 80 ml of 6% acrylamide mixture with 450 μl of 10% ammonium persulfate and 120 μl of TEMED. Pour immediately into the clamped gel mould. Insert a 0.4 mm comb between the plates at the top of the gel. Let the gel set for 30–60 min and remove the comb carefully.
3. Pre-electrophorese the gel for 30–60 min at 40 W.

Protocol 3. *Continued*

4. Mix 4 μl of labelled PCR product with 4 μl of formamide dye.
5. Denature the DNA sample in a boiling water-bath for 5 min and then put on ice for 2–5 min.
6. Flush the wells of the 6% polyacrylamide gel to remove any excess urea.
7. Load 6 μl of the mixture into each well.
8. Electrophorese using 0.5 × TBE running buffer at 40 W for about 2 h, until the light blue dye reaches the bottom of the gel.

Protocol 4. Autoradiography

Equipment and reagents

- 10% acetic acid
- Whatman 3MM paper
- Gel drier (e.g. Bio-Rad)
- X-ray film (e.g. Kodak XAR)
- Cassette for autoradiography (without intensifying screens)

Method

1. Fix the gel in 10% acetic acid for 20–30 min.
2. Place a sheet of 3MM Whatman on the top of the gel, and remove any air bubbles.
3. Rinse the gel with tap-water for 20 min.
4. Dry the 3MM Whatman sheet by blotting it with a layer of paper towels with a heavy weight on top. Leave for 5 min and remove the towels.
5. Remove the paper to which the gel is attached from the glass plate. Cover the gel with Saran wrap.
6. Dry the gel at 80°C under vacuum for 20 min.
7. Place the dried gel in a cassette with an X-ray film on top, and leave at −80°C.
8. Develop the film after one to three days.

4.5 Interpretation and analysis of the gel

Allele sizes are determined on the autoradiograph according to the M13 sequencing ladder, which is used as a molecular size standard. Alleles that have been sized in this way can then themselves be used as size markers on other gels. Size markers should be loaded in the middle of the gel, and in the lanes on either end to avoid sizing errors from differences in migration across the gel.

In theory, each allele of a target microsatellite sequence should give rise to a single band. However, extra bands (stutter) with a two nucleotide spacing arise as a result of slipped-strand mispairing during the PCR (11). Homozygotes therefore produce two or more bands and heterozygotes at least four bands. The true alleles, however, can be distinguished from the stutter bands, since the former are more intense. The interpretation is a little more difficult in an individual who is heterozygous for two alleles which differ by only one copy of the repeat. In this case the stutter band of the larger allele (A1) co-migrates with the real band from the smaller allele (A2), and results in a band of greater intensity in the A2 position. This pattern can easily be distinguished from a homozygote for allele A1, since the lower stutter bands will be considerably less intense than the top band. The degree of stuttering varies from one CA repeat to another, and can generally be reduced to some extent by a reduction in the number of PCR cycles.

The CA strand of the amplified DNA fragment migrates faster than TG strand. The same strand should therefore be labelled in every experiment with a particular marker. It should also be remembered that CA repeats have mutation rates of about 5×10^{-4} per gamete, so that new alleles which are not present in the parental genotypes will occasionally be observed.

5. Fluorescent labelling and detection

5.1 Marker selection and choice of fluorescent label

The basis of marker selection is essentially the same as that described in section 4.1. Once suitable markers have been selected, some planning is required in the choice of label colour for each marker in order to make maximum use of the polychromatic fluorescent system. There are currently four dye colours available. One of these (Rox: red) is used for labelling the molecular size standards, while the remaining three are used to label the microsatellite markers. The list below is an outline of the approach we have found most useful.

(a) Label the markers which have non-overlapping allele size ranges (minimum gap of 15 bases) with a dye of the same colour.
(b) Label up to three markers which have overlapping allele size ranges with dyes of a different colour.
(c) Redesign the primers for markers which have overlapping allele size ranges.

The size of the amplified product of a MSP can be increased or reduced by redesigning the primers with sequence data from the genome database (GDB) so that more loci can be analysed in a single gel lane. More than 2000 human microsatellite markers are now available from GDB, thus providing a wide range of markers for each chromosome. Since the PCR products from most MSPs are between 80–400 bp in length, and have a mean allele size range of 20 bp, it should be possible to analyse 24–30 loci in a single gel lane. This represents a significant improvement on the maximum of four markers which can generally be analysed simultaneously by radionuclide labelling and autoradiography.

5.2 Fluorescent labelling, purification, and storage of primers

Oligonucleotide PCR primers can be obtained commercially and may be supplied already fluorescently labelled and purified (see Appendix). However for reasons of cost this is not always an option, and we routinely synthesize and label primers in the laboratory. The fluorescent dyes currently available for labelling primers are either of the NHS-ester type or the phosphoramidite type. The former require an aminohexyl linker to be added to the primer during synthesis followed by a separate labelling reaction, and the purification protocol must include a step to separate excess dye from the labelled primers. The phosphoramidite dyes are easier to use since they are attached directly to the primer during synthesis (see *Protocol 5*), and are purified in a single step.

We have restricted our protocols to the preparation of phosphoramidite labelled primers. The cost of labelling a 0.2 μM oligonucleotide synthesis is about 10% lower using the phosphoramidite dyes compared to the NHS-type dyes, although the yield from the former is slightly lower. However, once the phosphoramidite dye is resuspended in diluent its coupling efficiency falls off significantly after two days. It is therefore necessary to plan in advance the sequential synthesis of seven or eight oligonucleotides that can be labelled with the same colour dye. Phosphoramidite dyes are now available in three colours, 6-FAM, HEX, and TET.

Oligonucleotides are produced as a standard 0.2 μM synthesis on a DNA synthesizer, with some modifications for the synthesis of those which are labelled with phosphoramidite dye (see *Protocol 5*). Oligonucleotides are deprotected and cleaved from the support column in ammonium hydroxide (*Protocol 6*). Oligonucleotides which prime the forward reaction are synthesized with a fluorescent dye phosphoramidite at the final 5′ position, and are purified by passage through an oligo purification column (OPC) to remove unlabelled oligomers and failed sequences (*Protocol 7*). Oligonucleotides which prime the reverse reaction are unlabelled and require no further purification.

Protocol 5. Oligonucleotide synthesis with fluorescent labelling

Equipment and reagents

- DNA synthesizer and recommended reagents (e.g. Applied Biosystems ABI 391)
- 6-FAM, HEX, or TET phosphoramidite dye, 120 mg (Applied Biosystems)
- Acetonitrile diluent (Applied Biosystems)

Method

1. Set-up the DNA synthesizer for a standard 0.2 μM synthesis cycle.
2. Resuspend the phosphoramidite dye in 1 ml of acetonitrile (sufficient for seven to eight couplings) and attach the bottle to the synthesizer at position 5.
3. Program in the required oligonucleotide sequence with X at the extreme 5′ end.
4. Select the Trityl ON option.
5. Place the appropriate oligonucleotide column in position and start the synthesis.

Protocol 6. Elution and deprotection of oligonucleotides

Equipment and reagents

- Disposable 2 ml syringes
- Speed-vac vacuum centrifuge system, SS21 (Savant Inst. Inc)
- Ammonium hydroxide (30%)

Method

1. Insert a syringe in one end of the column containing the oligonucleotide and insert into the other end another syringe containing 1 ml of ammonium hydroxide (NH_4OH).
2. Cycle the NH_4OH back and forth through the column six times taking care not to introduce bubbles into the column.
3. Leave at room temperature for 45 min, and then repeat step 2.
4. Leave at room temperature for a further 1.5 h, and then cycle the NH_4OH through again.
5. Draw all the fluid into one syringe and transfer to a 1.5 ml screw-top tube.
6. Seal the tube with Parafilm and use a waterproof label. Incubate at 55°C in a water-bath for 3.5 h (fluorescently labelled oligos) or overnight (unlabelled oligos).
7. Divide the preparation into four tubes. Remove the caps and centrifuge under vacuum until all of the ammonium hydroxide has evaporated, i.e. until completely dry.[a] Store at 4°C.

[a] Drying time depends on the initial total volume in the centrifuge but also on the volume per tube. Usually 12 tubes of 250 μl dry in 2.5 h.

Protocol 7. Purification of fluorescently labelled primers

Equipment and reagents

- Oligonucleotide purification column (OPC, Applied Biosystems)
- Vacuum centrifuge
- 2 M triethylammonium acetate (TEAA, Applied Biosystems)
- Acetonitrile, 8% in 0.1 M TEAA
- 10 ml disposable syringes
- Acetonitrile, 100%
- Acetonitrile, 20% in water

Method

1. Remove the plunger from a 10 ml syringe and attach the syringe to one end of the OPC. Add 5 ml of acetonitrile to the syringe and pass through the column, to waste, using gentle pressure from the plunger.
2. Pass 5 ml of 2 M TEAA through the OPC to waste.
3. Resuspend one tube of the oligonucleotide preparation (from *Protocol 6*, step 7) in 1 ml of 0.1 M TEAA. Pass the solution slowly through the OPC. Collect the eluate and pass it through the OPC a second time.
4. Pass 5 ml of 8% acetonitrile, in 0.1 M TEAA, through the OPC to waste. Repeat this step once.
5. Pass 5 ml of water through the OPC to waste.
6. Elute the oligonucleotide slowly by passing 0.5–0.7 ml of 20% acetonitrile in water through the column. Collect the eluate in a 1.5 ml screw-top tube.
7. Regenerate the OPC by repeating steps 1 and 2.
8. Purify the second oligonucleotide preparation tube by repeating steps 3–6.
9. Purify the remaining two tubes by repeating steps 1–8 with a new OPC.
10. Vacuum centrifuge the total eluate (four tubes of 0.5 ml 20% acetonitrile) until dry. Store at 4°C.

Protocol 8. Estimation of yield, and storage of PCR primers

Equipment and reagents

- Speed-vac vacuum centrifuge system, SS21 (Savant Inst. Inc)
- 0.01 M triethylammonium acetate (TEAA, Applied Biosystems)
- T(0.1)E buffer

Method

1. Resuspend one tube (25%) of the oligonucleotide preparation in the appropriate buffer. For unlabelled primers (from *Protocol 6*, step 7) this

is 200 μl of T(0.1)E. For labelled primers (from *Protocol 7*, step 10) this is 100 μl of 0.01 M TEAA.

2. Prepare suitable dilutions (usually 1 in 200) and estimate the yield from the OD at 260 nm. (1 OD = 33 μg/ml of single-stranded oligomers.)
3. Calculate the volume containing two nanomoles (for 20 bases 6.67 μg = 1 nM) and prepare six aliquots of this volume.
4. To one aliquot add T(0.1)E buffer to 200 μl and label. This is a working solution of 10 pM/μl. Store at 4°C.
5. Remove the caps and vacuum centrifuge the remaining aliquots until dry.
6. Label and store all tubes of dried oligonucleotides at −20°C.

Approximate yields from a 0.2 μM oligonucleotide synthesis, after labelling and purification, are as follows: unlabelled primers, 1.2 mg; 6-FAM/HEX amidite labelled primers, 400–500 μg. The final yield of fluorescently labelled primer from a 0.2 μM synthesis is sufficient for 12 000–18 000 amplification reactions. A single oligonucleotide synthesis should therefore generate sufficient primers for many different linkage studies. We have stored stocks of labelled primers at −20°C for up to 12 months and working solutions at −4°C for up to four months. All primers remain stable and give good fluorescent signals.

5.3 Establishing conditions for multiplex PCR

As for radionuclide labelled PCR products (see section 4), fluorescently labelled multiplex PCRs are best designed on a trial and error basis to decide which markers are compatible in a given set of reaction conditions. However, since many fluorescently labelled markers can be electrophoresed together this can be a complex process. We recommend the system described in *Protocol 9* in order to minimize the time involved.

Protocol 9. Multiplex PCR for fluorescent analysis

Equipment and reagents

- DNA thermal cycling machine
- Minigel apparatus
- Polyacrylamide (non-denaturing) 5% gel
- 1 × TBE buffer (*Protocol 3*)

Method

1. For each pair of primers perform single PCR using the most convenient reaction conditions (use *Protocol 1* or *Protocol 10*).
2. Separate the products on a 5% polyacrylamide minigel and visualize by ethidium bromide and UV light.

Protocol 9. *Continued*

3. Repeat any reactions that fail, altering one parameter of the PCR at each trial, until optimum conditions are found. Alternatively, discard these primers and select a new marker.
4. Combine up to eight pairs of primers (which will amplify in the same reaction conditions) in a single 5 μl volume PCR using *Protocol 1* or *10*.
5. Separate the products by electrophoresis on the Genescan fluorescent system (see *Protocol 11*).
6. If any primers fail to amplify or give only a weak signal increase the primer concentration. If non-specific bands are produced then decrease the primer concentration.
7. If the problem in step 6 is not resolved, amplify those particular primer pairs separately and pool an equal volume of the PCR products.
8. Each multiplex or pooled PCR should now give fluorescent signals of roughly equal intensities (see text below). Separate the multiplexed/pooled PCR products on the Genescan system (see *Protocol 11*), loading aliquots of 0.5–3.0 μl per gel lane to establish which volume gives the optimal signal for all the bands.

Protocol 10. PCR conditions for amplification of fluorescent microsatellites

Equipment and reagents

- DNA thermal cycler
- T(0.1)E buffer
- 10 × *Taq* polymerase buffer: 67 mM Tris base pH 8.8, 166 mM $(NH_4)_2SO_4$, 1.7 mg/ml BSA, 12.5 mM $MgCl_2$
- 5 mM dNTP solution: 5 mM each of dCTP, dATP, dTTP, dGTP
- Forward (labelled) and reverse (unlabelled) PCR primers 10 pM/μl[a]
- *Taq* DNA polymerase (5 U/μl)
- Mineral oil

Method

1. Transfer 1 μl of each DNA (50 ng)[a] into separate 0.5 μl reaction tubes.
2. Mix (for every ten reactions):
   - 32.2 μl T(0.1)E
   - 5.5 μl *Taq* polymerase buffer (10 ×)
   - 2.2 μl dNTP (5 mM)
   - 1.7 μl of each primer (10 pM)
   - 0.7 μl *Taq* DNA polymerase

   Vortex and spin briefly.

3. Add 4 μl of the mix to each DNA in the reaction tubes.
4. Overlay each reaction mixture with mineral oil.
5. Denature the reaction mixture at 94°C for 10 min. Then amplify with 24 cycles[a] of:
   - 94°C/0.75 min
   - 55°C/0.75 min
   - 72°C/2.0 min

   Include a final extension step of 72°C/10 min.

[a] Parameters which can be varied if the PCR fails or gives non-specific bands.

In multiplex PCR, variation occurs in the quality of results from each pair of primers. Non-specific bands or multiple stutter bands have been seen on the fluorescent gels which were not apparent on the trial agarose gels, and it is therefore necessary to check the amplification products on the fluorescent system. The amount of PCR product from the different loci, and thus the fluorescent signal strength (peak height) will also vary. For example, in a multiplex of five markers on chromosome 7, the peak heights obtained from ten DNA samples varied from 75 to 1311 units. The Genescan software allows adjustment of the data scale, from 8000 to 25 units, so that bands of high and low intensity can still be scored. However, when the template DNA quality is poor, the markers with the weakest intensity of fluorescence in a multiplex are not seen at all and require individual amplification. An alternative to spending time on establishing conditions for multiplex PCRs is to amplify loci individually, and to pool the PCR products prior to loading a gel. However not only does this technique use up the limited stock of patient DNA, but it also increases the consumption of DNA polymerase, and is generally considered to be tedious work.

5.4 Electrophoresis and analysis of gels

Multiplexed or pooled fluorescently labelled PCR products are separated by electrophoresis (see *Protocol 11*) using an automated DNA fragment analyser (Applied Biosystems). This protocol applies to the standard (24 cm) gels run on the 373 Fragment Analyser. A new instrument, the 373 Stretch DNA sequencer, can be used to run 12 cm gels. These provide adequate resolution for microsatellites, and can be run faster, thus providing increased throughput.

The data is collected automatically and transferred to a gel collection file on the Macintosh computer. Data stored in each gel collection file is then pre-processed and analysed using the Genescan 672, version 1.2 software. The pre-process and analysis parameters are optimized to suit each gel run. Detailed advice about the use of this system and the software is given in the manuals. However, an outline of the sequence of steps is given in *Protocol 12*.

Protocol 12. Polyacrylamide gel electrophoresis

Equipment and reagents

- 373A DNA Fragment Analyser (Applied Biosystems) and dedicated Macintosh computer, with 8 MB RAM (Apple)
- Loading buffer: deionized formamide with dextran blue
- DNA size standard, labelled with red (Rox) dye, 37–1181 bp (Applied Biosystems)
- 6% denaturing polyacrylamide gel (*Protocol 3*, but filter before use)
- 10 × TBE buffer (*Protocol 3*, but filter before use)

Method

1. Mix fluorescently labelled PCR products (0.5–3.0 μl) with an equal volume of loading buffer and add 0.5 μl of the size standard.
2. Denature samples at 94°C for 3–6 min, then transfer tubes immediately to ice.
3. Load samples on to a denaturing 6% polyacrylamide gel and run the electrophoresis in 1 × TBE buffer.
4. Run the gel at constant power (40 W) for 4–7 h.[a]

[a] Length of run depends on the size of the amplified products, fragments of 300–400 bases are resolved after 6–7 h at this power.

Protocol 12. Pre-processing and analysis of fluorescent gels

Equipment

- Macintosh computer, Centrus 650 or equivalent, uprated to 8 MB RAM (Apple)
- Genescan 672 software, version 1.2

Method

1. Create a sample sheet containing information about which DNA sample has been loaded in each gel lane.
2. Adjust pre-process parameters to compensate for:
   - length of the gel run
   - sensitivity of the fluorescent signals
3. Select the sample sheet created in step 1. Pre-process the collection file. A gel picture will appear on the computer screen.
4. Adjust the background and intensity values for each colour in order to show each marker band clearly (see *Figure 2*).

5. Select the analysis parameters. These allow:
   - compensation for the type of gel (polyacrylamide or agarose) used
   - adjustment of the sensitivity, so that weak signals can be analysed
   - a choice of statistical method for size calling
6. Select the gel lanes and colours to be analysed and the type and colour of the size standards.
7. Analyse the lanes. The results are then presented in the form of an electrophoretogram (see *Figure 3*) and spreadsheet.

The electrophoretogram format is easy to interpret and the distinction between true alleles and the 'stutter' bands is facilitated by comparison of peak height, with 'stutter' bands being weaker in intensity than allele bands. The 'stutters' do not cause an interference problem as long as markers labelled with the same colour have more than 10 bp between their nearest allele bands. By leaving a 15 base gap (see section 5.1) between such allele size ranges we allow for the possibility of allele bands which are outside of the published range.

The results are also presented in a spreadsheet format containing the size, height, and location of each peak in the electrophoretogram. The estimation of allele size is achieved by reference to the internal size standards, which should compensate for any lane to lane variation. The 672 Genescan software allows a choice of statistical methods for size calling. We use the least squares method (second order) which is based on regression analysis, and should result in the minimum error for all the fragments. In our hands, this method gives consistent results within families on a single gel run and for individual DNA samples run on different gels. Allele sizes are often observed to vary from published values by one base. In the protocols described here the forward reaction primer is the one which is labelled with the fluorescent dye. The majority of publications on the identification of microsatellites do not specify which strand (CA or GT) was used to determine allele size, therefore the discrepancy between published and observed values may be the result of differences in the mobility of the two strands. These size differences, although small, could result in significant variation in linkage calculations if the population frequencies of alleles that differed by one repeat unit were very different.

5.5 Automated reading of results

Results can be read automatically using the ABI Genotyper software package. The Results file is exported into Genotyper, and appropriate parameters such as peak height and the expected allele size range are then selected. Minor, extraneous peaks are then filtered out automatically. This data is tabulated, and provides the genotype of each marker for every DNA sample in the sample sheet. The data is then in a form in which it can be exported

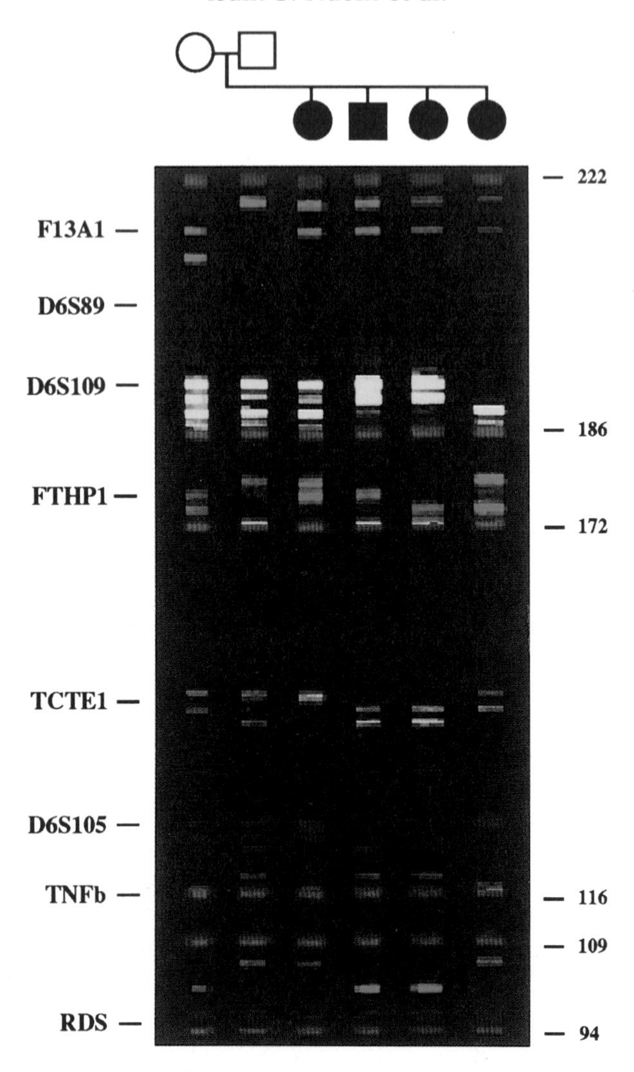

Figure 2. A Genescan gel display showing a family typed with a multiplex of eight microsatellite marker loci on the short arm of chromosome 6. The markers are labelled with either yellow (TAMRA), green (HEX), or blue (6-FAM). Internal size standards are shown in red (ROX), and the sizes are indicated on the right of the gel in base pairs.

into other files such as a spreadsheet (e.g. Excel), and then transferred directly into a linkage analysis program.

5.6 Data collection and storage

In addition to the fragment analyser and dedicated computer, further items

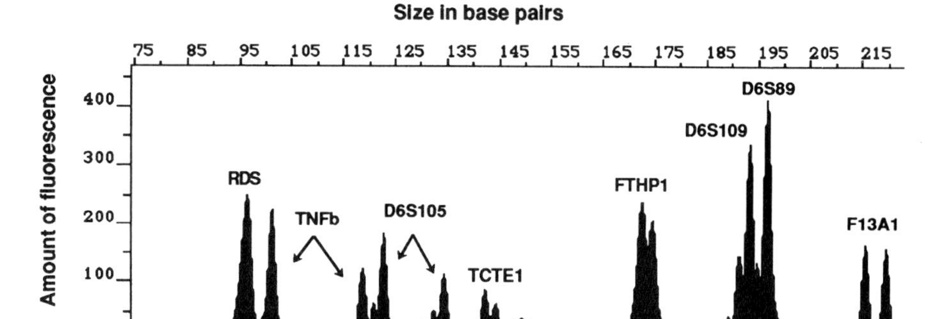

Figure 3. Electrophoretogram display (in black and white) of a multiplex PCR assay of the eight markers on chromosome 6p, taken from lane 5 of *Figure 2*. Each peak represents an allele, and is displayed on screen in the colour with which the primer is labelled. The molecular sizes of alleles are calculated automatically according to the internal size standards.

of hardware are required (see list below) in order to adequately process and store the large amount of data generated during a linkage study.

(a) Macintosh computer, Centrus 650, or equivalent uprated to 8 MB of RAM (Apple).
(b) A network link between the Macintosh computers.
(c) Colour printer, to obtain a hard copy of the results. Ink-jet type printers are suitable for electrophoretograms, but printing gel pictures will require a more expensive laser printer.
(d) Optical disc drive, 650 MB (Apple) and optical discs (d2/Sony).

The second computer is required as a workstation to which data can be transferred via a network. In this way one gel can be processed and analysed while the next is being run. A printer can be regarded as optional, particularly when data is stored on discs, but it is very useful for the preparation of presentations and reports. Storage of data is only possible by means of the optical discs, since each processed and analysed gel from a run of 6–7 h occupies approximately 15 MB of memory, and will rapidly use up the available space on the hard drive of the computer. Although these items add significantly to the cost of the equipment, they are necessary when the system is in constant use.

6. Conclusion

The ubiquity of microsatellite polymorphisms, their high degree of polymorphism, and the fact that they can be typed by PCR, has led to their complete dominance of mammalian genetic maps within five years of their discovery. Our enthusiasm for these markers should be tempered with a few words of caution. First, many of the markers based on dinucleotide repeat polymorphisms produce 'stutter' bands of a length two, four, and six nucleotides shorter than the genuine alleles (see section 3.2). These extra bands are usually less intense than the full-length product, but some care and experience is required to interpret the band patterns produced. Secondly, DNA polymorphisms involving a single base change can occur within the binding site of one of the PCR primers used to amplify the MSP, which may lead to a failure of amplification of an allele at relatively high annealing temperatures. It is therefore advisable to type a new MSP in a panel of reference families to check for non-Mendelian behaviour before it is used, for example, for diagnostic purposes. Finally, the relatively high mutation rate of MSPs, which has generated such diversity of alleles in the population, will also lead occasionally to the detection of a novel allele in an individual which is not present in either of his or her parents. Typing such a family with several other MSPs will soon establish whether you are dealing with a new mutation or some other problem such as non-paternity or a sample mix-up. However, these are minor caveats compared to the enormous harvest of genetic information which will be reaped from microsatellites within the next few years.

Acknowledgements

We thank Iftekhar Haris for introducing us to the Applied Biosystems automated DNA sequencer. This work is supported by grants from the Medical Research Council, the Robert McAlpine Trust, the National Association of Colitis and Crohn's disease, and the Generation Trust.

References

1. Saiki, R., Gelfand, D. H., Stoffel, S., Scharf, S. J., Higuchi, R., Horn, G. T., *et al.* (1988). *Science*, **239**, 487.
2. Weber, J. L. and May, P. E. (1989). *Am. J. Hum. Genet.*, **44**, 388.
3. Litt, M. and Luty, J. A. (1989). *Am. J. Hum. Genet.*, **44**, 397.
4. Smeets, H. J., Brunner, H. G., Ropers, H. H., and Weiringa, B. (1989). *Hum. Genet.*, **83**, 245.
5. Ott, J. (1986). In *Human genetic diseases: a practical approach* (ed. K. Davies), pp. 19–32. IRL Press, Oxford.
6. Gyapay, G., Morisette, J., Vignal, A., Dib, C., Fizames, C., Millasseau, P., *et al.* (1994). *Nature Genet.*, **7**, 246.
7. Dietrich, W. F., Miller, J. C., Steen, R. G., Merchant, M., Damron, D., Nahf, R., *et al.* (1994). *Nature Genet.* **7**, 220.
8. Serikawa, T., Kuramoto, T., Hilbert, P., Mori, M., Yamada, J., Dubay, C. J., *et al.* (1992). *Genetics,* **131**, 701.

9. Zheng, L., Collins, F. H., Kumar, V., and Kafatos, F. C. (1993). *Science*, **261**, 605.
10. Cornall, R. J., Prins, J. B., Todd, J. A., Pressey, A., de Larato, N. H., Wicke, L. S., *et al.* (1991). *Nature*, **353**, 262.
11. Jacob, H. J., Lindpaintner, K., Lincoln, S. E., Kusumi, K., Bunker, R. K., Mao, Y. P., *et al.* (1991). *Cell*, **67**, 213.
12. Farrall, M. (1991). In *Protocols in human molecular genetics* (ed. C. Mathew), Vol. **9**, pp. 365–88. Humana Press, Clifton, New Jersey.
13. Lander, E. S. (1988). In *Genome analysis: a practical approach* (ed. K. Davies), pp. 171–89. IRL Press, Oxford.
14. Litt, M. (1992). In *PCR: a practical approach* (ed. M. J. McPherson, P. Quirke, and G. R. Taylor), pp. 85–99. IRL Press, Oxford.
15. Yuille, M. A. R., Goudie, D. R., Affara, N. A., and Ferguson-Smith, M. A. (1991). *Nucleic Acids Res.* **19**, 1950.
16. Edwards, A., Civitello, A., Hammond, H. A., and Caskey, C. T. (1991). *Am. J. Hum. Genet.*, **49**, 746.
17. Love, J. M., Knight, A. M., McAleer, M. A., and Todd, J. A. (1990). *Nucleic Acids Res.*, **18**, 4123.
18. Ziegle, J. S., Su, Y., Corcoran, K. P., Nie, L., Mayrand, P. E., Hoff, L. B., *et al.* (1992). *Genomics*, **14**, 1026.

Appendix

1. The genome database (GDB) contains human genetic mapping information on genes, clinical phenotypes, and DNA segments arranged by chromosomal location. It includes a listing of all known polymorphisms, primer sequences, map location, and references.
   To access GDB, contact:
   GDB User Support,
   Welch Medical Library,
   1830 E. Monument St., 3rd floor,
   Baltimore, MD 21205, USA.
   Telephone: USA 401–955–9705
   Fax: USA 401–955–0054
   E-mail: help@welch.jhu.edu
   Several European nodes of GDB are also available. Details can be obtained from GDB.
2. Unlabelled and fluorescently labelled oligonucleotide primers can be purchased from:
   Oswel,
   Department of Chemistry,
   University of Edinburgh,
   Kings Buildings, West Mains Road,
   Edinburgh EH9 3JJ.
   Scotland, UK.
   Fax: (031) 662 4054
   For suppliers in other countries, contact Applied Biosystems, Foster City, California, USA (fax 415–570–6667) for further information.

A1

Addresses of suppliers

Advanced Protein Products, Unit 18H, Premier Partnership Estate, Leys Road, Brockmoor, Brierley Hill, West Midlands DY5 3UP, UK.

American Type Culture Collection, 12301 Parklawn Drive, Rockville, MD 20852, USA.

Amersham International plc, Lincoln Place, Green End, Aylesbury, Buckinghamshire HP20 2TP, UK; 2636 Sth Clearbrook Drive, Arlington Heights, IL 60005, USA.

Amicon, Amicon Division, W. R. Grace and Co., 72 Cherry Hill, Beverley, MA 01915, USA.

AMS Biotechnology (UK) Ltd, 5 Thorney Leys Park, Witney, Oxon OX8 3DN, UK.

Anderman and Co. Ltd, 145 London Road, Kingston-Upon-Thames, Surrey KT17 7NH, UK.

B. Braun Biotech, 999 Postal Road, Allentown, PA 18103, USA.

Bayer Diagnostics, Evans House, Hamilton Close, Basingstoke, Hampshire RG21 2YE, UK.

Beckman Ltd, Progress Road, Sands Industrial Estate, High Wycombe, Buckinghamshire HP12 4JL, UK; 1050 Page Mill Road, Palo Alto, CA 94304, USA.

Becton Dickinson UK Ltd, Between Towns Road, Cowley, Oxford OX4 3LY, UK; Labware, 2 Bridgewater Lane, Lincoln Park, NJ 07035, USA.

Bio-Rad Laboratories Ltd, Bio-Rad House, Maylands Avenue, Hemel Hempstead, Hertfordshire HP2 7TD, UK; Alfred Nobel Drive, Hercules, CA 94547, USA.

Boehringer-Mannheim UK, Bell Lane, Lewes, East Sussex BN7 1LG, UK; Biochemical Products, 9115 Hague Road, PO Box 50414, Indianapolis, IN 46250–0414, USA.

Branson Ultrasonics Corporation, 41 Eagle Road, Danbury, CT, USA.

British Biotechnology Products Ltd, 4–10 The Quadrant, Barton Lane, Abingdon, Oxon OX14 3YS, UK.

Collaborative Medical Products Inc, Becton Dickinson Hardware, Two Oak Park, Bedford, MA 01730, USA.

Corning Inc, Science Products, MP-21-5-8, Corning, NY 14831, USA.

Difco Laboratories Ltd, PO Box 13B, Central Avenue, West Molesley, Surrey KT8 0SE, UK; PO Box 1058, Detroit, MI 48232, USA.

Dupont/NEN Research Products, 549 Albany Street, Boston, MA 02118, USA.

Falcon, (distributed in USA by Becton Dickinson Labware).

Gibco Life Technologies, 8400 Helgerman Court, Gaithersburg, MD 20877, USA.

Heat Systems Inc, 1938 New Hwy, Farmingdale, NY, USA.

Hybaid, 111–113 Waldegrave Road, Teddington, Middlesex TW11 8LL, UK; National Labnet Corporation, PO Box 841, Woodbridge, NJ 07095, USA.

ICN Biomedicals Inc, 3300 Hyland Avenue, Costa Mesa, CA 92626, USA.

International Biotechnologies Inc, (IBI Limited, A Kodak Company), 36 Clifton Road, Cambridge CB1 4ZR, UK; PO Box 9558, 25 Science Park, New Haven, CT 06535, USA.

Invitrogen Corporation, (distributed in UK by British Biotechnology Products Ltd); 3985 B Sorrento Valley Building, San Diego, CA 92121, USA.

Life Technologies, PO Box 35, Trident House, Renfrew Road, Paisley PA3 4FF, UK.

Merck-BDH, Merck Ltd, Poole, Dorset BH15 1TD, UK.

Millipore Corporation, 80 Ashby Road, Bedford, MA 01730, USA.

New England Biolabs/C.P. Laboratories, PO Box 22, Bishop's Stortford, Hertfordshire CM23 3DH, UK; 32 Tozer Road, Beverley, MA 01915–5599, USA.

Novagen Ltd, (see AMS Biotechnology (UK) Ltd as UK distributor); 597 Science Drive, Madison, WI 53711, USA.

Nunc A/S, PO Box, Kamstrup, DK-4000, Roskilde, Denmark.

Perkin-Elmer Cetus, 761 Main Avenue, Norwalk, CT 06856, USA.

Pharmacia, Davy Avenue, Knowlhill, Milton Keynes MK5 8PH, UK; PO Box 1327, Piscataway, NY 08855–1327, USA.

Pharmacia LKB Biotech Ltd, 23 Grosvenor Road, St. Albans, Hertfordshire AL1 3AW, UK; 800 Centennial Avenue, PO Box 1327, Piscataway, NJ 08854–1327, USA.

Promega Ltd, Delta House, Enterprise Road, Chilworth Research Centre, Southampton SO1 7NS, UK; 2800 Woods Hollow Road, Madison, WI 53711–5399, USA.

QIAGEN Inc, (distributed in UK by Hybaid); 9259 Eton Avenue, Chatsworth, CA 91311, USA.

Sarstedt Ltd, 68 Boston Road, Beaumont Leys, Leicester LE4 1AW, UK.

Schleicher and Schuell, (distributed in UK by Anderman and Co. Ltd); 10 Optical Avenue, PO Box 2012, Keene, NH 03431, USA.

Sigma Chemical Company Ltd, Fancy Road, Poole, Dorset BH17 7TG, UK; PO Box 14508, St. Louis, MO 63178, USA.

Sorvall Du Pont Ltd, Wedgewood Way, Stevenage, Hertfordshire SG1 4QN, UK; PO Box 80024, Wilmington, DE 19880–0024, USA.

Sterogene Biochemicals, San Rafael, California, USA.
Stratagene Ltd, Cambridge Innovation Centre, Cambridge Science Park, Milton Road, Cambridge CB4 4GF, UK; 11099 North Torrey Pines Road, La Jolla, CA 92037, USA.
Vector Laboratories Inc, 30 Ingold Road, Burlingame, CA 94010, USA.

Index